T0179967

Understanding Race

The human species is very young, but in a short time it has acquired some striking, if biologically superficial, variations across the planet. As this book shows, however, none of those biological variations can be understood in terms of discrete races, which do not actually exist as definable entities. Starting with a consideration of evolution and the mechanisms of diversification in nature, this book moves to an examination of attitudes to human variation throughout history, showing that it was only with the advent of slavery that considerations of human variation became politicized. It then embarks on a consideration of how racial classifications have been applied to genomic studies, demonstrating how individualized genomics is a much more effective approach to clinical treatments. It also shows how racial stratification does nothing to help us understand the phenomenon of human variation, at either the genomic or physical levels.

Rob DeSalle is a curator in the Sackler Institute for Comparative Biology and the Program for Microbial Research of the American Museum of Natural History, New York, USA. His research focuses on molecular systematics, microbial evolution, and genomics. He is the author of over 500 scientific papers and a wide range of books, from popular science titles to textbooks on genomics.

Ian Tattersall is Curator Emeritus in the Division of Anthropology of the American Museum of Natural History, New York, USA. His most recent research is on the emergence of modern human cognition. He is author of over 400 scientific papers and numerous books, is a prominent interpreter of palaeoanthropology to the public, and writes regularly for *Natural History*.

The *Understanding Life* series is for anyone wanting an engaging and concise way into a key biological topic. Offering a multidisciplinary perspective, these accessible guides address common misconceptions and misunderstandings in a thoughtful way to help stimulate debate and encourage a more in-depth understanding. Written by leading thinkers in each field, these books are for anyone wanting an expert overview that will enable clearer thinking on each topic.

Series Editor: Kostas Kampourakis http://kampourakis.com

Published titles:

Understanding Evolution	Kostas Kampourakis	9781108746083
Understanding Coronavirus	Raul Rabadan	9781108826716
Understanding Development	Alessandro Minelli	9781108799232
Understanding Evo-Devo	Wallace Arthur	9781108819466
Understanding Genes	Kostas Kampourakis	9781108812825
Understanding DNA Ancestry	Sheldon Krimsky	9781108816038
Understanding Intelligence	Ken Richardson	9781108940368
Understanding Metaphors in the Life Sciences	Andrew S. Reynolds	9781108940498
Understanding Cancer	Robin Hesketh	9781009005999
Understanding How Science Explains the World	Kevin McCain	9781108995504
Understanding Race	Rob DeSalle and Ian Tattersall	9781009055581

Forthcoming:

Understanding Human Metabolism	Keith N. Frayn	9781009108522
Understanding Fertility	Gab Kovacs	9781009054164
Understanding Human Evolution	Ian Tattersall	9781009101998
Understanding Forensic DNA	Suzanne Bell and John M. Butler	9781009044011
Understanding Natural Selection	Michael Ruse	9781009088329
Understanding Creationism	Glenn Branch	9781108927505
Understanding Species	John S. Wilkins	9781108987196
Understanding the Nature–Nurture Debate	Eric Turkheimer	9781108958165

Understanding Race

ROB DESALLE
American Museum of Natural History, New York

IAN TATTERSALL
American Museum of Natural History, New York

WITH ILLUSTRATIONS FROM PATRICIA WYNNE

CAMBRIDGE
UNIVERSITY PRESS

University Printing House, Cambridge CB2 8BS, United Kingdom

One Liberty Plaza, 20th Floor, New York, NY 10006, USA

477 Williamstown Road, Port Melbourne, VIC 3207, Australia

314–321, 3rd Floor, Plot 3, Splendor Forum, Jasola District Centre,
New Delhi – 110025, India

103 Penang Road, #05–06/07, Visioncrest Commercial, Singapore 238467

Cambridge University Press is part of the University of Cambridge.

It furthers the University's mission by disseminating knowledge in the pursuit of
education, learning, and research at the highest international levels of excellence.

www.cambridge.org
Information on this title: www.cambridge.org/9781316511374
DOI: 10.1017/9781009052450

© Rob DeSalle and Ian Tattersall 2022

First published 2022

Printed in the United Kingdom by TJ Books Limited, Padstow Cornwall

A catalogue record for this publication is available from the British Library.

ISBN 978-1-316-51137-4 Hardback
ISBN 978-1-009-05558-1 Paperback

"DeSalle and Tattersall provide a brilliant and comprehensive refutation of the folk concept of human races. Anyone who thinks that there are natural categories of people that correspond to zoological subspecies will have their worldview blown to bits!"

Jonathan Marks, Department of Anthropology,
University of North Carolina at Charlotte

"*Understanding Race* explains to the reader in accessible terms all the misconceptions that continue to plague both lay people and professionals concerning race. First the authors establish for the reader the fundamental mechanisms of evolution that are responsible for the variation within all species; then they explain how people thought about variation before there was a science to correctly explain it. The book guides the reader through how racial thinking changed as our understanding of evolution, as well as the technology to understand genetic variation, improved. The authors end by drawing attention to ongoing misconceptions concerning biological variation and social definitions of race in a variety of arenas, including medicine. If you don't read my books, you should read theirs; and in the best of all worlds you should read both."

Joseph L. Graves Jr., Professor of Biological Sciences,
North Carolina A&T State University

To the Memory of Richard Lewontin
(1929–2021)
Who brought sanity to the issue of genetics and race

Contents

Foreword

The term "race" is unfortunately one that immediately brings to mind situations of discrimination and inequalities when it is applied to humans. But this is not due to any inherent differences between the human groups that are often distinguished from one another as being different races; rather, it is due to biases – conscious or unconscious – that make people think that internally homogeneous human races exist that are in turn clearly distinct from one another. The situation becomes even more complicated when attempts are made to naturalize these distinctions and divisions by explaining the existence of distinct human races on biological/genetic grounds. As Robert DeSalle and Ian Tattersall explain in this magnificent book, this is far from accurate from a scientific point of view. All available evidence, especially from human genomics, supports the conclusion that human genetic variation is continuous, not clustered. This means that the genetic variation of any two human groups is overlapping, and that we share most of our DNA. Of course, there are average differences between human groups, but these differences do not support the division of humans in genetically distinct human races, or groups of any kind. Does this mean, then, that human races do not exist? The answer that the authors give is that if they do, they depend on culture and not biology. If people decide to differentiate themselves from outgroups in favor of their ingroup, this is something done by choice or culture and upbringing, and is not imposed by our genetic background. DeSalle and Tattersall invite you on a fabulous journey that presents the scientific evidence for the fact that all humans living today are members of a huge family that has evolved very recently in evolutionary terms. Reading this book will make you appreciate how much we share in common, and wonder why we often insist on paying attention to our very minor differences.

Kostas Kampourakis, Series Editor

Preface

Race matters. Historically, economically, and culturally, race matters a lot. In the United States, for example, a straight and uninterrupted line of distress can be drawn between slavery in the eighteenth and nineteenth centuries and the mass incarcerations of African Americans in the twenty-first. A similar line connects the early nineteenth-century miseries of the "Trail of Tears" (a series of horrendously long forced marches in which members of various southeastern indigenous American tribes were made to relocate to unfamiliar new western territories, with at least one-quarter of them dying of disease and exhaustion along the way) to the conditions of deep deprivation that prevail on many Native American reservations today. These important historical factors cannot be ignored; and without accommodating them we cannot explain, or understand, or even begin to improve the deeply flawed social world we live in. And there is equally no doubt that those historical and current travesties are inextricably intertwined with notions of race.

Yet we will argue in this book that, biologically speaking, human races do not exist. That is because, to be accessible to science, something must not only be observable but also definable and preferably measurable. This in turn requires that the entity observed possess recognizable boundaries. And, as we will see, human races simply do not meet this criterion at the biological level. Certainly, striking variations in hereditary physical features can be observed between members of human populations originating in different areas of our planet, just as similar variation exists among the individuals who compose those populations. But on closer examination the boundaries between those populations fade, not just because most human physical features vary in a continuous manner, but because the human varieties – the races – we perceive are constructs not of Nature, but of the human mind. Beyond the human mind, it turns out, races have no objective biological existence.

Of course, perhaps the most remarkable uniqueness of we human beings is that we live – at least for much of the time – in the worlds we construct in our minds, rather than in the world as Nature directly presents it to us. Our artificial constructions of the world are of critical importance to the quality and conduct of our daily lives; and indeed, we couldn't live our complex interior existences without them. But this makes it all the more important for us to understand when the distinctions that are filtered through our mental processes are accurate reflections of Nature itself, and when they are by-products of the ways in which we arbitrarily represent the world. If we are to remain grounded in reality, it is essential to ensure that the images we entertain of the world around us are as accurate as possible, and are not simply products of our preconceptions, or objects of our convenience.

That essential grounding in reality is what science is there to provide; and nowhere is scientific understanding more important than in the all-pervading question of race. We hope to show in this short book that, while our notions of "race" are based on the undeniable reality of human biological variation – and yes, in some respects human beings do indeed differ strikingly, if not significantly – they are in fact artifacts of subjective human perception that vanish under closer scrutiny. And since, like all other organisms, we are the product of a long and complicated evolutionary history, we will begin by looking at the evolutionary background.

But before we do that, let's just note that words matter too. As members of a species that uses language to communicate, we have developed into a curious group of organisms whose words can be angry, soothing, creative, influential, and, sadly, viciously fatal. Our words thus need to be precise in order to avoid misunderstanding. The one word that we focus on in this book – "race" – has been used to propagate some of the most evil, pernicious, and regrettable episodes in the history of human existence, which makes it necessary to be precise about what this word means, and how we use it. A casual reader might conclude that we spend too much time on interpretation, potentially reducing the problems we discuss here to ones of semantics; but that would be wrong. Outside its use for humans, the word "race" has a precise meaning in science; and it is that scientific meaning, and how that meaning is implemented by scientists, that we address in detail because human beings are, of course, an integral part of the natural world. Race also, of course, has distinct but somewhat less precise definitions in the cultural and social spheres; and it is our purpose here to dissect the exact scientific meaning of the word and to show that it does not apply to humans, in order to place those less precise social and cultural uses of the word in scientific context.

Acknowledgments

We thank Kostas Kampourakis, editor of this unique series of books, for his kind invitation to write this volume, and for his astute comments on our initial draft. At Cambridge University Press, our particular appreciation goes not only to Katrina Halliday, the editor under whom this project started, and to Jessica Papworth, the editor under whose guidance it was completed, but also to Olivia Boult who kept everything flowing, to Sam Fearnley, Akash Datchinamurthy and Felinda Sharmal who oversaw production, to Gary Smith for his sensitive copy-edit, and to Judith Reading for the index. And, as ever, we are immensely indebted to our wonderful artist, Patricia Wynne. Thank you all; it has been great working with you.

1 The Evolutionary Background

How Evolution Works

Like every one of the many millions of other organisms with which we share our planet, the species *Homo sapiens* is the product of a long evolutionary history. The first very simple cellular organisms spontaneously arose on Earth close to four billion years ago, and their descendants have since diversified to give us forms as different as streptococci, roses, sponges, anteaters, and ourselves. That evolution was responsible for that variety was first formally articulated in 1858 by the British natural historians Charles Darwin and Alfred Russel Wallace, and the nature of the process(es) involved has been vigorously debated ever since (see the companion volume in this series, *Understanding Evolution*, by Kostas Kampourakis). As exhaustively documented by Darwin in his 1859 masterpiece *On the Origin of Species by Natural Selection*, the nested patterns of physical similarity that we see among organisms are best explained by a long history of branching descent from a succession of common ancestors: The essence of evolution is, as Darwin neatly put it, "descent with modification." Living species, in other words, give rise to descendant species that are not exactly like them; and, with the passage of enough time, any initial differences between parental and offspring species will become hugely amplified. Predictably enough, this radical view of a dynamically changing biota initially raised a huge furor in the static mid-nineteenth-century Creationist world into which it exploded; but once the immediate fuss had died down, the notion that life had evolved was quite rapidly accepted by most Victorian scientists. Even the conservative British public and the ecclesiastical authorities were, by and large, not too far behind. What continued to be controversial was not the notion of evolution itself, but the process that Darwin put forward to explain it.

That process was "natural selection." Darwin spent a lot of time studying the activities of animal breeders (he raised pigeons himself) and he saw how, by choosing which individuals were allowed to reproduce, those breeders could induce changes from one generation to the next in the physical characteristics of the animal stocks they wished to improve. He thus came to believe that a similar process of selection could happen naturally, and that it could help him to overcome the greatest obstacle he faced in convincing his fellow scientists and others that evolutionary relatedness was the explanation for the order seen in Nature: the received wisdom, enshrined in the Scriptures, that the living species on the planet now are just the same as when the Creator had placed them there. In short, Darwin saw that in each species individuals varied in their physical characteristics – an unarguable quality without which domestic animal breeding could never have worked. It thus logically followed that some individuals would be better suited – better adapted – to prevailing environmental circumstances than others were, enhancing not only their survival but also their ability to reproduce. And if those favorable adaptations were inherited – a reality that breeders had depended on from time immemorial, even if they didn't have a clue how that inheritance worked – those fortunate individuals would pass those features along to their offspring. Provided that the environmental pressures remained constant, such blind and entirely situational reproduction bias would produce changes in the population over generations, as the favorable variations inexorably became more common; and, given enough time, a new species could emerge.

Humans are storytelling creatures, and natural selection proved to be an amazingly compelling story. Legend has it that Darwin's colleague Thomas Henry Huxley slapped his forehead and exclaimed, "how stupid of me not to have thought of that!" when he heard about natural selection for the first time. Still, without knowing just how inheritance worked, the natural selection story was obviously incomplete; and it was not until the elucidation in 1900 of the principles of heredity that most of us associate with the Czech cleric Gregor Mendel that this gap began to be filled. Historian Robert Olby was the first to point out that Mendel's 1866 conclusions were not identical to those that are generally attributed to him; but whenever biologists speak of "Mendelian" genetics they are referring to the fundamental notions that individual traits are controlled independently (of each other) by discrete hereditary elements that are passed on intact through the generations, and that remain discrete even though "dominant" alternative forms might mask the effects of "recessive" ones in the phenotype. And Mendel certainly deserves credit for rebuffing the vague assumption, prevalent in his day, that parental characteristics become somehow "blended" in the offspring.

In 1883 the German biologist August Weismann determined that in multicellular organisms the hereditary material is transmitted solely in the female ovum and the male sperm, a finding that opened the way for the independent (re)discovery of Mendelian genetic principles in three separate European laboratories at the turn of the century. In turn, this quickly led to the introduction of the term "mutation" for the spontaneous changes in the hereditary material that provide the potential for evolutionary change, and to the application of the term "gene" to the particulate unit of heredity that might mutate. There then followed a period of hyperactivity, by the end of which it had become generally recognized that most physical traits of individuals were determined not by single genes, as those studied by Mendel had conveniently been, but by many of them. It was also rapidly realized that the environment also played an important role in the determination of phenotypes. As early as 1915, the American geneticist Thomas Hunt Morgan and his colleagues had shown that at least 25 genes were responsible for eye color in the fruit fly *Drosophila*, and that an alteration in only one of them could bring about phenotypic change. By 1918 such considerations had led the statistician and population geneticist Ronald Fisher to his "infinitesimal model" of heredity, which saw most phenotypes (physical features of the individual) as the outcome of an interplay between the environment and large numbers of different genes, each of which was individually of minor effect. By admitting environmental agency, this model opened the door for natural selection to re-enter the scene as a potential agent of evolutionary change, a role that was solidified with the emergence during the 1920s and 1930s of what became known as the "Modern Evolutionary Synthesis." This emerging consensus united geneticists, systematists, and paleontologists in seeing evolutionary change as principally driven by the gradual action of natural selection over vast periods of time: Species evolved as their members slowly adjusted under its influence to changing environments, or as they became better adapted to the ones they occupied. Applying this stately process to any evolutionary history would naturally transform it into a linear story of constant improvement, which rapidly became received wisdom about the process itself.

The reductionist appeal of the Synthesis was extremely seductive, but it was eventually forced to yield to empirical observations of the fossil record that suggested an absence of the linear evolutionary patterns it predicted. In the early 1970s, paleontologists began pointing out that instead of gradually transforming themselves out of existence as predicted by the Synthesis, well-known fossil species tended typically to linger in the record for extended periods, before being abruptly replaced, often by close relatives. This observation was hardly surprising in the light of a flood of paleoenvironmental data that increasingly revealed that climatic and biotic changes in the past had tended to occur rather suddenly, and with alarming frequency. Ancient environments had typically oscillated on timescales so short that it was difficult to imagine natural selection keeping up as a force for adaptive change. Certainly, the kind of steady, long-term environmental change that one might see as driving natural selection over the necessary timescales seems to have been rare, at best.

Realizations such as these opened the door to a new view of evolution as a complex and multifaceted process into which many influences intruded, not least among them random chance. Geneticists had long recognized the potential effect on evolutionary histories of what they called "drift," namely, the incorporation of particular alleles (alternative versions of given genes) into small populations for reasons of nothing more than random chance (just as if you put equal numbers of black and white marbles in a bottle, you will nonetheless have a decent chance of shaking out of it two black ones in a row). What is more, new alleles arise spontaneously, by mutations that regularly occur without regard to their effects on the individuals in which they take place. We will return later to the nature of those mutations; for now, it is enough to note that they may be of three kinds: significantly disadvantageous, in which case there is a good chance that they will disappear from the population; significantly advantageous, which gives them a good chance of being conserved; and basically neutral, in which case they may disappear but may equally stay around if they don't actively get in the way. If you add random elements of the last kind to the inherent instabilities of the environments in which populations live (which have the effect of periodically breaking up larger ones into the smaller fragments most susceptible to drift), then it becomes clear that evolution cannot be the burnishing process, always tending toward improvement, that the Synthesis promoted. Evolution is clearly not about optimization, as is often assumed. Rather, any population's fate on the evolutionary stage is a result of its practical ability to get by (or not) with whatever Nature has dealt it, both in terms of its own inherent qualities and those of the environment (which includes any competition there may be around, as well as the resources available).

All of this puts natural selection in a new light. For, while selection is a mathematical certainty in any population in which more individuals are born than survive to reproduce (which is to say, all of them), its major role may in fact lie in stabilizing gene frequencies in populations, and thereby keeping those populations viable over time. That is something it achieves not by promoting directional change, but by trimming off the unfavorable extremes (say, individual vertebrates with two heads, or none). Indeed, it is hard to see how the gradual modification of many traits simultaneously might be achieved. For a start, individuals rarely succeed or fail in the reproductive stakes as the result of one or two favorable characteristics. That is because every individual is an incredibly complex integrated whole, and needs to function as such. It's not ordinarily of much advantage, for example, to be the strongest or most fecund member of your species or society if you are also the slowest or the most short-sighted. From a survival point of view, especially in a highly social species, it is usually much better to be in the middle of the pack. And whether you survive or not is often a matter of random chance: An unfortunate encounter with a predator, a falling tree branch, or even a slip of the foot can easily carry away even the most magnificently endowed individuals. What's more, most features of any individual are controlled developmentally by many genes, each of which may be involved in turn with the development of many characteristics, so that changing one allele will probably have ramifying effects, and every change will involve a trade-off. It is thus very hard to see how natural selection can home in on one feature to favor over the ages, except, perhaps, in the rare event that the feature involved is absolutely key to survival or reproduction.

What this means is that we need to get away from the idea that populations within species are composed of individuals all engaged in constant competition to be the most optimized in some respect or another. Rather, both those individuals and the populations they belong to are reasonably characterized as grab-bags of features and alleles that have historically proven good enough to get by, and many of which are present in them purely as random occurrences. As we'll see, this certainly appears to be the case in *Homo sapiens*.

Species and Subspecies

Nature is wonderfully diverse. And the most fundamental unit within that mind-boggling diversity is the species. Our species is *Homo sapiens*, the first component of this two-part "binomen" being the name of our genus, *Homo*, and the second designating our species in particular. Genera are the larger groups into which related species are assembled; and in obeisance to the nested pattern of resemblances produced by the branching evolutionary process, genera are in turn grouped into subfamilies, which are grouped into families, which belong to orders, and so on up until all living things are subsumed. Just for the record, a unit at any level of this hierarchy can be referred to as a "taxon" (plural "taxa"), hence the term "taxonomy" for the study of natural diversity.

The intuitive recognition of species as the basic unit into which Nature is organized goes back a long way, perhaps as far back as words themselves. And in some respects, species recognition has remained intuitive. In 1859 Charles Darwin wrote that "No one definition [of species] has satisfied all naturalists; yet every naturalist knows vaguely what he means when he speaks of a species." That sage declaration still rings true: There are currently on offer almost three dozen different formal definitions of what sexually reproducing species are. What most of those modern definitions have in common, though, is that they mostly recognize those species less by what their members look like than by their ability to interbreed with each other and produce fully viable and fertile offspring. In other words, most biologists would nowadays concur that, among sexually reproducing organisms, species are the largest fully interbreeding populations – an idea whose origins go back as far as the seventeenth-century speculations of the English natural historian John Ray.

Unfortunately, that doesn't get us too far in practice, hence the plethora of definitions, most of which apply to specific circumstances. That's largely because the inability to reproduce successfully (i.e., to produce fully viable and fertile offspring), which is what confirms that two populations have gone off on their own independent paths, can express itself in many ways. A handful of acts of interbreeding between individuals from two populations, for example, does not guarantee the production of viable and fully fertile offspring, and does not necessarily mean that the parental populations of both individuals would reintegrate if they came into contact. In fact, it may well be that what is or is not a species is only determinable in long retrospect, because any reproductively isolated new species inevitably starts with the splitting of a single species, the members of which may continue sending out reproductive signals and interbreeding should the opportunity arise, even after having embarked on an individuated evolutionary trajectory. This is what appears to have happened in the case of *Homo sapiens* and *H. neanderthalensis*, two very physically distinctive species of the genus *Homo* that initially went their separate evolutionary ways (in Africa and Eurasia, respectively) well over half a million years ago. The two apparently interbred on numerous occasions once *H. sapiens* had eventually reached the Neanderthals' European heartland after leaving Africa at around 70,000 years ago. The resulting encounters resulted in the exchange of various genes between the two species (most of them later weeded out of the gene pools of both), but they did nothing to affect the evolutionary fate of either party in any really significant way. The Neanderthals became extinct pretty much as they were, while *H. sapiens* went on to conquer the rest of the world as the entity we know today (with high-altitude populations on the Tibetan Plateau apparently receiving a genetic assist from a mysterious Neanderthal relative known currently only as the Denisovans). At around the same time, all other remaining archaic hominin populations went extinct as well, presumably because they, too, were unable to cope with competition from our species.

In this book we fortunately do not have to deal with the larger questions of human identity raised by complications such as these coexistences. Because, as a result of the late Ice Age purges of the competition, all modern humans belong to the single freely interbreeding species *Homo sapiens*. There is no question about that. What is more, as we will see later, most of the variations we see today in the extraordinarily dense and widespread populations of our species are no more than epiphenomena of the last 70,000 years or so, and none of them can predate its origin in Africa at about 200,000 years ago. And because our forebears eliminated all human competition on our planet a pretty long time ago, we modern *Homo sapiens* are not faced with having to philosophize about who is human, and who is not. We all are. The relevant question is: Is our young but widespread species currently divisible in any meaningful way?

Old-time zoologists used to recognize species, the basic "kinds" of organisms, on the basis of physical similarity. Doing this raised all kinds of issues since, as Charles Darwin was one of the first to make explicit, all species consist of individuals that vary among themselves. You might be able to demonstrate that a measured physical characteristic varies around a mean value (a value that might not be observed in even one individual in the sample available); but there is certainly no single individual out there that, in every particular, represents the "ideal" of the species. That such an ideal exists is a tenacious notion that has hung around since Aristotle, but it is obviously problematic since any member of a species can be accused of falling short in some respect – but remains a member, nonetheless.

Still, the "population thinking" that Darwin advocated raised issues as well, because as more was learned about the behaviors of nonhuman animals in their natural environments, the plainer it became that populations of closely related organisms that appeared different to the human eye sometimes – even often – could, and would, interbreed if given the chance. From this there eventually emerged the idea that among most animals, including mammals, distinctive features and reproductive incompatibilities evolved in small populations in isolation, the implication being that if a species was divided into discrete populations by the formation of some impassable geographic boundary such as a major river, over time they would eventually accumulate enough genetic differences to become reproductively incompatible with their former conspecifics. But even as those reproductive incompatibilities accumulated, behavioral attraction might remain among those close relatives, so that if the populations somehow reestablished physical contact, some interbreeding behaviors might take place. If effective speciation – the establishment of effective reproductive incompatibility at any level – had occurred, that behavioral integration would continue only until the greater reproductive success of those who declined to interbreed edged out any inclinations to do so in the population at large. In this perspective, for all their physical differences, Neanderthals and modern humans first encountered each other before behavioral isolating mechanisms had completely evolved. But it is the bottom line that is critical, and the Neanderthals are nonetheless long gone.

The hominin case is a fairly extreme one in terms of the extent of the substantial physical differences between the quasi-reproductively compatible taxa involved. Still, it is nevertheless pretty typical among living species to exist as a number of readily recognizable varieties that inhabit particular geographic areas, but that appear to hybridize readily if and when they come into physical contact. In diurnal vertebrate species, especially, such varieties are often strikingly distinguished by the markings and coloration of their pelage, and they are often allocated to their own "subspecies." Subspecies are designated by a "trinomen," with a third italicized name added after the species name. In the human case this was done, for example, by those who at one time believed that the extinct Neanderthals should be included in the modern human species, as *Homo sapiens neanderthalensis*. That opened a huge can of worms, because if the Neanderthal variety of humankind should formally be recognized as a subspecies, shouldn't the other "kinds" of humans be similarly distinguished? Fortunately, it has turned out that the Neanderthals amply justify recognition as their own species, so the issue is moot. Subspecies are taxa that are readily recognized from their physical and often genomic characteristics, but that are nonetheless genetically wholly compatible reproductively with members of other populations of the same species. When you are deciding whether it is worth making a subspecies distinction in a particular case, it is important to remember that the differences that cause taxonomists to recognize different subspecies are wholly maintained by their possessors' physical separation. They will rapidly disappear, through interbreeding, if and when geographical circumstances allow the populations to merge. If they don't merge under those circumstances, you are not looking at subspecies, but instead at differentiated species.

So, while subspecies, having emerged in isolation, undoubtedly provide the fodder for new species (without them, it is difficult to know how existing species could split to form new evolutionary lineages to create the diversity of species we see), we have also to remember that they are by nature ephemeral, with an entirely transient existence until they are historically established as discrete units by speciation – the establishment of that key reproductive isolation that allows them to embark upon independent historical existences. Crucially, this means that we are unable to recognize subspecies in the species *Homo sapiens*. That's because it could hardly be more evident that the many geographical variants within it are busily intermingling right now, worldwide. And, as we will see, they have a long history of doing so.

Human Evolution

At around the middle of the twentieth century, the Modern Evolutionary Synthesis found its way into paleoanthropology, the branch of science devoted to the evolution and diversification of human beings and their close relatives. The new view of this process brought with it an elegantly simple and linear story of human phylogeny that had the appeal of all the best stories from time immemorial, featuring as it did a single protagonist that, with the aid of natural selection, single-mindedly battled its way across the eons from the status of primitive bipedal ape to that of the perfected product that some believe *Homo sapiens* is today. The triumph of this interpretation brought with it a welcome clearing-out of the clutter of names that had accumulated for the denizens of what was then a fairly limited fossil record. But subsequent discoveries have changed the picture once more; and although the seductiveness of the linear story still shows significantly in the desire of many paleoanthropologists to minimize the diversity of species and genera they are prepared to recognize in the human fossil record, almost three-quarters of a century down the line it is hard for anyone to deny that the actual story of human evolution involved a lot of trial and error, rather than the burnishing of a single central lineage.

In a book about human beings and race, it is important to point out that, while subspecies may be quite readily distinguishable among living animals on the basis of external features, it is usually difficult or impossible to spot them if you only have the bones and teeth that typically fossilize. In many cases it is tricky even to identify fully differentiated fossil species. Looking in the human fossil record for taxa below the level of the species is, therefore, likely to be a pretty unproductive endeavor, although in the literature you may still spot the occasional trinomen (and, in one spectacular instance, a quadrinomen: *Homo erectus ergaster georgicus*, for what it's worth). In the last decade or two the availability of genomic information has made it possible to talk about population movements of early *Homo sapiens* subsequent to about 30,000 years ago; but as far as extinct kinds of human are concerned (with the partial exception of the Neanderthals), all the traditional paleontologist can say is that, although we can often clearly observe a lot of variation in bony features among individuals that may or may not have belonged to the same species, we have no way of judging if, let alone how, they were differentiated below that level. What the human fossil record does give us, though, is evidence that in the past several different kinds of hominin typically occupied the planet – and sometimes the very same landscape – at the same time. The situation we naturally enough take for granted – that *Homo sapiens* is the only hominin around – is, in fact, highly unusual. So unusual, in fact, that our current lonely splendor is very probably entirely unprecedented in the seven million years since the first member of our subfamily – the first hominin – took to walking upright on *terra firma*. This means that, in our own case at least, we need to be cautious in taking the past as a guide to the present.

The hominin subfamily (hominid family, if you prefer – for our purposes, the difference is notional) was born in Africa at a time when an increasing aridity and seasonality of rainfall was transforming the continent's environment. Especially to the east of the elevated Great Rift Valley that runs in a ragged line from Djibouti to Mozambique, the formerly ubiquitous humid African forests were drying out and giving way to more open woodlands and bushy formations. All this put pressure on the ancestral apes that inhabited those forests, while at the same time presenting new ecological opportunities on the ground; and in the period between about seven and four million years ago we have African fossil evidence of several different early hominin genera, all of which appear to have walked erect on the ground. As far as we can tell, they all also shared another peculiarity with us: the reduction of the canine teeth that has important consequences for what we can chew, and thus eat. Whether all – or even any – of these forms were directly ancestral to later hominins can be debated. What is clear, though, is that the open terrestrial environment offered valuable resources to any apes willing to spend significant amounts of time away from the trees; and that some of those apes, at least, were already accustomed to holding their trunks upright because they were highly suspensory climbers in the trees. Their preexisting upright proclivity would then have translated to the ground – as it does, for example, among today's gibbons. Aside from the immediate challenges of earthbound existence, such as figuring out how to dig out the underground tubers that were to be a major component of their diets, these terrestrial pioneers also faced some pretty difficult realities: Moving on two legs was a slow way of getting around, and the new environment teemed with fast and ferocious predators.

At around 4.4 million years ago we find the first evidence of a much better-documented group of early hominins known colloquially as the "australopiths" (for the best-known genus of the group, *Australopithecus*, which translates as "southern ape"). Some australopiths persisted until well under two million years ago (see Figure 1.1, which shows the diversity of hominins over time). As Figure 1.1 shows, a dozen australopith species are known from fossils found in eastern, central, and southern Africa, and all have in common short stature, brains not a lot bigger than those of apes, projecting faces with large chewing teeth, and lower bodies showing a commitment to bipedality when on the ground. They retained climbing adaptations in their upper bodies, however, and appear to have divided their lives between the trees and the ground. Nonetheless, it was almost certainly an australopith that made the first stone tool at some time over about 2.5 million years ago – probably the most consequential invention ever in hominid history. Hominids were therefore already stone-tool makers well before the time when the first species of our genus *Homo* emerged in Africa, at some point just under two million years ago. The ancestry of this species, commonly known as *Homo ergaster*, probably lay somewhere within the larger australopith group, though specific claims about this ancestry are probably better treated with caution.

Homo ergaster – taller and slenderer than any australopith, and significantly larger-brained – initially made stone tools like those made by the australopiths: simple sharp flakes obtained by bashing one fist-sized cobble with another. But soon they were making "handaxes," larger stone tools that were made to a specific (usually teardrop) shape that must have existed in the toolmaker's mind before the work began. These hominids were also the first that were committed by their anatomy (much like our own) to life away from

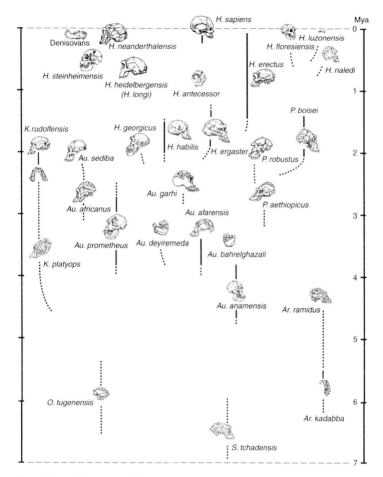

Figure 1.1   Highly provisional hominin family tree, showing the longevities of the species involved and some possible relationships. Note that for virtually all of the subfamily's existence several different hominin species have been present at each time point.

the trees: They were "obligate" rather than "facultative" bipeds, meaning that they were dedicated to walking upright rather than just being able to do so as the occasion demanded. They are known to have used their stone tools to butcher the animal carcasses that would have been necessary to feed their energy-hungry large brains (though it is an inference that they actively hunted them), and it has been convincingly argued that they would have needed to have cooked animal and plant parts – obviously using domesticated fire – to extract that energy. In the two-to-one-million-year period we have plenty of evidence for a diversity of "early *Homo*" hominids around, suggesting that active experimentation was going on with the potential inherent in belonging to *Homo*; but we have nothing that would help us to understand how their populations were organized biologically. As an aside, it is worth noting that some paleoanthropologists still classify African *H. ergaster* as *H. erectus*, even though it is becoming ever more apparent that the latter, despite its historical importance as the second species of extinct hominin to be discovered, was a much later and exclusively eastern Asian descendant of *H. ergaster*.

It was a million years or more after the invention of the handaxe that the next major stone tool type was invented, this time the "prepared-core" tool that required a stone "nucleus" to be laboriously shaped until a final blow would detach a more-or-less finished tool, with a continuous cutting edge around its periphery. This development took place within the tenure of *Homo heidelbergensis*, the first cosmopolitan hominin species we know of, with fossil representatives in Africa, Europe, and eastern Asia. *Homo heidelbergensis* had a brain not too far short of modern size, and it showed a whole panoply of complex behaviors, such as hunting with spears, the construction of artificial shelters, and the fabrication of compound tools with stone points mounted in wood or bone handles. But it exhibited few, if any, inklings in its recorded behaviors of the unusual way in which modern humans handle information in the mind. We are "symbolic," which is to say that we deconstruct the world around and within us into a vocabulary of discrete mental symbols that we can shuffle around, according to rules, to imagine alternatives. We do not simply live in the world that Nature presents directly to us: We live for much of the time in the worlds we create in our minds, and much of the discord we observe in society results from the fact that we each do this in our own idiosyncratic way.

Interestingly, we only arguably find evidence for symbolism in the complex and extensive record left behind by *Homo neanderthalensis*, a close human cousin that had a lineage in Europe extending back to over 400,000 years, and a brain as large as that of contemporaneous *Homo sapiens*. This latter fact suggests that our mental peculiarities result from a unique internal organization of the brain, rather than from raw brain size; and indeed, the earliest *Homo sapiens* fossils, known from Ethiopia at about 200,000 years ago, left no evidence of substantive cognitive difference from the Neanderthals (and, suggestively, modern human brains got smaller once the symbolic way of processing information had been adopted). Only at around 100,000 years ago do we begin to find indications (again, in Africa) that humans were beginning to think symbolically. Those indications come mainly in the form of objects of bodily decoration and geometric engravings that evidently held meaning for their makers; and since such objects appear significantly *after* the event of biological reorganization that gave rise to the very anatomically distinctive species *Homo sapiens*, the conclusion must be that a new biological potential acquired in that reorganization had been finally released by some behavioral factor, most plausibly the spontaneous emergence of spoken language. Such an event would certainly not have been evolutionarily unprecedented, or even unusual: The dinosaur ancestors of birds, for example, possessed feathers for many tens of millions of years before co-opting them for flight. And once those rudimentary symbolic expressions had begun to appear, the blossoming *Homo sapiens* rapidly left Africa and took over the entire world, displacing all other hominins in the process – and, in short order, producing some of the most powerful artistic expressions ever made, on the cave walls of France and Spain (and now of Sulawesi and Borneo).

The adoption of symbolic thinking changed the entire relationship of human beings to the world around them, and it quite probably changed the rules of the evolutionary game as we play it. But it probably didn't immediately affect the way in which human populations were organized. Like the members of any other species, local human population nuclei were buffeted hither and yon by the climatic vicissitudes of the late Ice Ages, fragmented at one moment by drought, and reunited at another when the rains returned. All along, hominins had been hunters and gatherers, necessarily living off whatever bounty the landscape offered, even after their new cognitive powers had allowed them to exploit with unprecedented intensity the environments that they happened to find themselves in. At certain points they were limited to tiny refuges that could sustain them even in the hardest times, and it was probably in one such isolated place that the anatomically recognizable species *Homo sapiens* originated in one local population, as explained in greater detail in the companion volume in this series, *Understanding Human Evolution*. That population would have been small enough to allow the fixation of whatever genomic change it was that had resulted in the highly unusual new anatomy (modern humans are much slenderer and more lightly built than even their closest known fossil relatives, and have radically reorganized skull proportions). The fledgling species was then poised to acquire symbolism (again, in one small subpopulation), and subsequently to spread throughout Africa and beyond. It is hard to know exactly what to make of the various fossils from around the Old World that document this diaspora; and fortunately, as we will see, that story is much better told by the molecular record.

2 Race before Evolutionary Theory

The Ancient World

Humans have been roaming around the planet since the very beginning, encountering other humans in the process and doubtless forming opinions about them. And since human beings seem to have an innate urge to classify everything – doing which, after all, lies at the core of our way of mentally organizing and understanding the world around us – there is little doubt that, from the earliest days, our symbolic forebears categorized each other in some way. Moreover, there does seem to be substance to the notion that there is a maximum number of people with whom any individual can have meaningful social interactions – that number is often given as around 150 – and that beyond this known core group there is a tendency to think of people, to a milder or more marked degree, as "other." The larger and simpler to categorize the group at issue becomes, the easier this distinction is both to make and to oversimplify. But while the basic human tendency to categorize lies at the base of racism, it evidently does not directly specify it.

Some of the earliest historical evidence we have of ancient interactions among peoples with distinctive physical attributes comes from ancient Egypt (third through first millennia BC), which was already an amazingly complex intermingling of populations with diverse geographic origins. Egyptian temple walls are liberally sprinkled with images of presumably paler-skinned Pharaohs vigorously smiting dark captives from Nubia, far up the Nile. The origin of such captives, often bound for slavery, is readily recognizable from their physical features; but those unfortunate captives were political adversaries caught up in hostilities between adjacent powers, and they are unlikely to have been the victims of systematic exploitation based on origin. Indeed, given that lower Egypt was a major crossroad of the ancient world, with constant streams of people coming from the north, south, east, and west, it seems very questionable that organized racism as we know it would have been a major feature of ancient Egyptian society.

For the ancient Greeks, race as we understand it today was equally a nonissue. The Greeks were familiar with, and judgmental about, peoples of many different ethnicities and geographical origins; but they were usually more concerned with variations in temperament than with physical appearance. What's more, the differences that were noted were generally regarded as shallow: The historian Herodotus recounted a Greek tendency to explain disparities in physical appearance among human populations (as well as differences in national character) as responses to the different environments in which they lived: to "airs, waters, and places," as he put it. And the general Greek disdain of the *barbaroi* was basically equal-opportunity: In Greek-controlled territories, non-Greeks of all origins encountered equal difficulty in achieving citizenship, or even full civil rights.

Finally, as violent, domineering, and ready to conscript anyone into slavery as the ancient Romans certainly were, they could only with the greatest difficulty be described as racists. As early as 300 BC the Romans had already begun granting full citizenship to any of their allies who wanted it, and eventually citizenship was extended to any nonenslaved resident within a vast empire that embraced a huge diversity of ethnicities and physiognomies. That didn't, of course, stop the metropolitan Romans from having a very strong sense of their own superiority; but the peoples they despised the most were their neighbors, precisely those who looked most like them. And while colonial people of African origin living in Rome were generally found among the less advantaged economic classes, in AD 193, after a bizarre and tortuous political episode, the Senate nominated Lucius Septimus Severus, a native of Libya, to be Emperor of the Roman Empire. Severus went on to become one of the most successful emperors, expanding the Empire to its greatest-ever area of almost two million square miles. The chances are that Severus himself was relatively light-skinned, but although the issue of his skin color has unfortunately been politicized in recent times, what does seem pretty evident is that, while the ancient world was undoubtedly rife with prejudices of various kinds (after all, it was populated by human beings), racism did not then exist in the way in which we know it today. As Julius Caesar had declared of the Empire in one of his Orations, delivered in the first century BC, "No one worthy of rule or trust remains an alien . . . all come together as into a common civic center, in order to receive each man his due."

All in all, it appears, then, that the particular brand of systemic racism that we are familiar with today is not an inevitable product of human perception. Rather, it is the product of a particular history.

Medieval and Post-Medieval Europe

After the Roman Empire crumbled in the fifth century AD, Europe generally looked inward. But Christianity and eventually the Crusades opened a portal to the east; then, as the Age of Exploration commenced in the fifteenth century, Europeans were reminded again that people existed in faraway lands who neither looked nor behaved exactly as they did. Indeed, they eagerly absorbed the tales told by explorers such as Antonio Pigafetta, one of the few survivors of Ferdinand Magellan's round-the-world expedition of 1519–1522, who claimed that Terra del Fuego was populated by giants twice as tall as he. Medieval bestiaries abounded with pictures not only of exotic animals that often existed only in the human imagination, but with strange versions of humanity that included the one-eyed Cyclops and the half-woman, half-bird Sirens.

Real people rapidly followed. It is usually supposed that the first sub-Saharan Africans to reach Britain in modern times were a group of five who arrived in 1554 from today's Ghana, brought to London to learn English by a trader who intended to use them as interpreters on future commercial voyages. But it is worth noting that one of eight skeletons of crewmen analyzed from the wreck of the warship *Mary Rose*, sunk in battle nine years earlier, was most likely that of an African. This is hardly surprising, because by that time the transatlantic slave trade from West Africa to the New World was already well underway, having been authorized by King Charles I of Spain as early as 1518. The first slave voyages involved mere handfuls of individuals, but by the mid-1520s the trade was already roaring, ships crammed to the gunwales with hundreds of miserable human beings.

Over the course of the next four centuries, more than 12 million Africans were transported directly from their homelands to slavery in the Caribbean and Brazil and beyond, with appalling loss of life. But all of this was happening far away from Europe; and, apart from substantially increasing local cash flow, this vile commerce had little direct impact on the European societies that profited from it. Those five Ghanaians were, indeed, politely received in London as curiosities, or even minor celebrities.

By the end of the sixteenth century, however, the welcome extended by Britain to increasing numbers of African immigrants (who might have arrived as slaves, but who had the status of free men as soon as they stepped off the boat) was wearing thin. In 1596, and again in 1601, Queen Elizabeth I attempted to expel allegedly freeloading Africans from England, reflecting an increasingly equivocal attitude on the part of the local populations, perhaps as the laboring classes began to feel threatened, something all too familiar today.

Given the role of the Church in European intellectual and political life, it is hardly surprising that the status of exotic (to Europeans) peoples should have been first seriously debated in the context of Christian theology. According to the Book of Genesis, all humans are ultimately descended from Adam and Eve, and therefore share a single common origin (the "monogenist" position). The diversity of humans in the world today results from the fact that, more proximally, Europeans, Africans, and Asians are respectively descended from Noah's three sons Shem, Ham, and Japhet – which still keeps everyone in the family. But by the early sixteenth century, and probably not coincidentally as the slave trade began to get underway, this orthodoxy was beginning to be questioned. In 1520 the Swiss physician and philosopher Paracelsus opined that people living in faraway places "did not descend from Adam," an opinion supported in the following century by the French theologian and lawyer Isaac de la Peyrère, who argued that "if Adam sinned . . . there must have been an Adamic law according to which he sinned . . . If law began with Adam, there must have been a lawless world before Adam, containing people." This suggestion of repetitive creations was not a big success with the ecclesiastical authorities, who had Isaac's books publicly burned, but it did reflect an increasing heterodoxy of clerical thought as it applied to the different geographical varieties of humankind.

In practical terms, the most consequential debate of this kind during the sixteenth century was without doubt the one that took place at the Vatican during the 1530s, concerning the status of the indigenous peoples of South America. The issue arose because even some pretty hard-bitten observers were becoming alarmed at the way in which those unfortunate populations were abused by their Portuguese and Spanish conquerors in the name of "Christianization." One side of the theological argument that followed was "polygenist," holding that the New World peoples had been created separately from Europeans, meaning that they had the status of beasts of burden, and had no inherent rights at all. On the other side of the debate were the monogenists, who felt that all people were brothers, with the rights due to all humans. Fortunately, the latter carried the day when, in 1537, Pope Paul III issued a papal bull in which he declared that the indigenous South Americans were "true men . . . capable of receiving the Christian faith . . . [and] must not be deprived of their freedom and the ownership of their property." This ruling was, alas, largely honored in the breach; and the situation rapidly deteriorated, to such an extent that by 1650 the aging Charles I felt obliged to convene a conference to decide whether or not colonization should continue. Eloquent presentations were made on both sides, but in the end a famously impassioned plea by the Dominican friar Bartolomé de las Casas that "all men are human" prevailed. Still, while colonization activities were legally suspended, the human and commercial plunder inevitably continued after a short pause.

Race in Enlightenment Thought

Secular thought was also affected by the explorations of the late fifteenth century and subsequently. Columbus had headed westward with a real fear of encountering some of the monstrous humans who populated those medieval bestiaries; but the discovery of the New World also gave rise to misty ideas of a sort of Eden in which people lived in harmony with Nature, and in the absence of the squalor and privation that were such unignorable features of European societies of the time. Although counterbalanced by the gloomier views of such writers as Thomas Hobbes, some seventeenth-century philosophers, most famously Jean-Jacques Rousseau (who never actually used the term), were entertaining notions of the "noble savage," a being untainted by the shortcomings of the present. As Rousseau put it, "we shall not conclude with Hobbes that just because he has no idea of goodness, man must be naturally wicked." Humans of this idealized kind had formerly existed in Europe, and still survived in distant corners of the world. The failure of native South Americans to develop iron-working technologies was, for example, taken as a badge not of primitiveness, but of a sort of pristine purity. At around the same time, however, the more pragmatic philosopher David Hume was able in 1748 to write, in pursuit of his search for "the constant and universal principles of human nature," that "Mankind are so much the same, in all times and places, that history informs us of nothing new or strange in this particular," a viewpoint that has in the long run proved much more durable. Yet Hume's fellow Scot Lord Kames (who had famously ruled in court that there could be no slavery in Scotland) was able to argue a quarter of a century later that the physical differences among human races were so great that climate, the factor to which they had been attributed at least since the Greeks, was insufficient to explain them. In a still devoutly Christian age, this secular conclusion necessarily hinted at polygeny.

A major underlying factor as Europeans struggled to incorporate human variety into their worldviews was a notion inherited from as far back as Aristotle. This was the Great Chain of Being, the idea that everything on Earth was ordered in a hierarchy from "lower" to "higher." This hierarchy started with inanimate rocks and minerals at the bottom, and proceeded inevitably upwards through pond scum to insects, through fish to mammals, to culminate eventually in humans at the top. This hierarchical arrangement had subsequently been seized on by medieval Scholastic theologians, who extended it ever upwards to include the angels and ultimately God. Every living thing had its place in this preordained hierarchy, and it eventually came to underpin the idea that the various geographical varieties of humankind could be ranked in a similar way. Significantly, however, the Great Chain of Being did not in the least color the first systematic attempt to organize human geographical variation. This was published in 1684 by the French physician François Bernier, under a title that translates as *A New Division of the Earth by the Different Species or Races Which Inhabit It* (Figure 2.1). Bernier had traveled widely in India and elsewhere, and he was a very perceptive and sympathetic social observer. He saw that Europeans, Africans, eastern Asians, and Laplanders were "races of men among which the difference is so

XII.   133

JOURNAL
DES SCAVANS,

ou

RECUEIL SUCCINT ET ABREGE' DE TOUT
ce qui arrive de plus surprenant dans la Nature, & de ce qui se fait
ou se découvre de plus curieux dans les Arts & dans les Sciences.

Du Lundy 24. Avril M. DC. LXXXIV.

NOUVELLE DIVISION DE LA TERRE,
par les differentes Especes ou Races d'hommes qui
l'habitent , envoyée par un fameux Voyageur à
M. l'Abbé de la * * * * * à peu prés en ces
termes.

LEs Geographes n'ont divisé jusqu'icy la
Terre que par les differens Pays ou Regions
qui s'y trouvent. Ce que j'ay remarqué dans les
hommes en tous mes longs & frequens Voyages,
m'a donné la pensée de la diviser autrement.
Car quoy que dans la forme exterieure du corps,
& principalement du visage, les hommes soient
presque tous differens les uns des autres, selon
les divers Cantons de Terre qu'ils habitent, de
sorte que ceux qui ont beaucoup voyagé peuvent
souvent sans se tromper distinguer par là chaque
nation en particulier ; j'ay neanmoins remarqué
qu'il y a sur tour quatre ou cinq Especes ou Races.
1684.   L l

Figure 2.1   The opening page of François Bernier's *Nouvelle Division de la Terre,* the first treatise on human variety around the globe. It was published in a 1684 issue of the *Journal des Scavans,* a learned periodical published in Paris that later changed its name to the *Journal des Savants.*

conspicuous that it can properly be used to make a distinction." But at the same time, he declared – more than half a century before Hume – that this differentiation, though real, involved little more than the colors of those people's skins. As far as he was concerned, all humans were cut from the same psychological and intellectual cloth, regardless of their physical features.

Human Races in Early Zoology

It was during the eighteenth century that zoology began to take on its modern shape, not least through the efforts of Carl Linnaeus. The Swedish natural historian clearly recognized the nested way in which living things are organized, and he introduced the system of classifying animals and plants we use today. The Linnean system includes the two-part species name (it was Linnaeus who baptized us *Homo sapiens* in 1758), and the grouping of genera into orders, and so on. Linnaeus's way of ordering Nature's diversity was thus hierarchical, but in a fundamentally different manner from the "exclusive" Great Chain of Being, in which each organism occupied a sole and unique place in the hierarchy of life. Linnaeus's scheme was "inclusive," so that each organism belonged not only to its own species, but to every rank that lay above it in the hierarchy. Bravely, Linnaeus included human beings in his classification of animals, placing them in the order Primates that also contained apes (to the extent then known), monkeys, and lemurs. He even went so far as to classify an ape in the same genus as human beings (as *Homo troglodytes*).

At the time, very little was known of the apes, apart from their uncanny physical resemblances to humans, and Linnaeus was probably giving them the benefit of the doubt. A similar kind of doubt was probably also involved in Linnaeus's failure to provide a physical diagnosis for the genus *Homo*: While each of the many other genera named in his great work, *Systema Naturae*, received its own thumbnail anatomical characterization, when it came to the group to which he himself belonged, Linnaeus did no more than dryly note "*nosce te ipsum*" (know thyself). Quite apart from the wisdom of avoiding an issue with which philosophers and others had wrestled for millennia before him (and continue to struggle with to this day), Linnaeus's discretion was probably also a nod to the ecclesiastical authorities with whom it was important for him to retain a cordial relationship. As long as he was just pointing out similarities, and didn't question the fixity of species, he was on theologically safe ground, merely clarifying the details of the Creator's design for Nature. As he somewhat smugly said, "God created; Linnaeus classified."

But while Linnaeus was able to perceive pretty clearly the characteristics that bound human beings to their close primate kin, he was also able to see a lot of differences among peoples from different parts of the world. Indeed, he was evidently so struck by those differences that in the definitive tenth edition of this *Systema*, published in 1758, he felt obliged to embrace them in his classification. He refrained from formally establishing a "subspecies" category for the varieties he recognized, but he nonetheless gave them names, and described the skin colors, body postures, temperaments (in good medieval fashion, according to the humors), distinctive traits, behaviors, and clothing styles of each. First listed was *"Ferus,"* (wild): four-footed, mute, and hairy. Linnaeus's account of this form apparently derived from the accounts of wild children, raised in the woods by animals, that abounded in his time). And last came *"Monstrosus,"* which embraced mythic forms such as Antonio Pigafetta's Fuegian giants.

The inclusion of these fantasies reminds us that in Linnaeus's day science as we know it was just getting underway, still as much bound by the past as it looked forward to the future. But besides his monstrosities Linnaeus also listed four readily recognizable geographical varieties of humans. First listed was *"Americanus,"* the "choleric" Amerindian group with red skins, cheerful temperaments, and governed by customary right. Next up was the "sanguine," white, and muscular *"Europaeus,"* inventive and governed by rites, followed by the sallow *"Asiaticus,"* melancholic, haughty, and governed by opinions. Finally, there was the African *"Afer,"* black-skinned, dark-haired, phlegmatic, sly, and governed by choice (by which he meant capricious). As our colleague Jon Marks has noted, all of these descriptions were consciously idealized: Linnaeus described Europeans "as having blue eyes and blond, flowing long hair . . . knowing full well that the vast majority of Europeans possessed neither of these features . . . he was describing what [Europeans were] *supposed* to look like." To which we might add that, especially for their day (and with the possible exception of *Afer*), these stereotypes are pretty unjudgmental.

Linnaeus was an obsessive classifier, a very different character from his almost exact French contemporary Georges Leclerc, Comte de Buffon. Buffon was a natural historian of encyclopedic interests, and he is still renowned as a freethinker who, while claiming to be a devout Christian, believed that all phenomena had natural causes. He saw no reason to believe in Noah's flood, and he derided theological estimates of the age of the Earth as far too short. His studies of vestigial organs in vertebrates led him to conclude that organisms could leave old functions behind them, even as they acquired new ones. Yet, even as he veered toward a notion of evolution in many of his observations, he declined to conclude that one species could indeed transmute into another. As he put it in the 1750s, a century before Darwin, "If it were once proved that [the pattern of Nature] could be established rationally ... there would be no limits to the power of Nature ... She could have drawn with time, all other organized beings from a single being ... But it is certain, from Revelation, that all ... species emerged fully formed from the hands of the Creator." From a modern standpoint that final qualification might seem like a politically convenient intellectual cop-out, a refusal to follow a train of thought to its logical conclusion. But it actually derived from Buffon's remarkably modern view of species as the basic bounded entity in Nature. Unlike Linnaeus, who wanted to characterize larger patterns in natural diversity by recognizing higher taxa, to Buffon the species was everything. There were no larger groupings. Still, he knew it was significant that species were variable, including the human one; and, like many of his contemporaries, he believed that species were capable of some change – just not into other species.

As word continued to arrive in Europe of humans of unfamiliar appearance living far away, Buffon (who was no fan of Linnean nomenclature) was quite ready to use the term "race" to denote the distinctive local human populations they represented. And he was happy to countenance that those races had histories. He recognized, for instance, that the close physical resemblances of the Amerindians to the inhabitants of Kamchatka strongly suggested that the Americas had been colonized from Siberia. In contrast to this picture of New World uniformity, Buffon noted the great "variety of men" in Africa, and suggested that this variety was very ancient. Those ideas would be entirely familiar to any modern molecular systematist, although in obeisance to tradition Buffon ascribed the differences concerned to climatic and other external factors. His overall conclusion was that:

> On the whole . . . mankind is not composed of species essentially different from each other . . . on the contrary, there was originally but one species which, after multiplying and spreading over the whole surface of the Earth, has undergone various changes by the influence of climate, food, mode of living, epidemic diseases, and the mixture of dissimilar individuals . . . at first these changes were not so conspicuous . . . [they] became afterwards . . . more strongly marked . . . and will gradually disappear . . . if the causes that produced them should cease.

Buffon was no evolutionist, but his viewpoint was astonishingly modern in seeing the human species as a variable and dynamically changing kaleidoscope of populations. His observations were compelling enough to make a deep impression on the philosopher Immanuel Kant, whose desire was to bridge the increasing gap between the empiricists who believed that experience is the sole source of knowledge, and the rationalists who believed that reason was the sole conduit to the truth. In biology, though, Kant tilted to the rationalist side. Impressed by Buffon's genealogical view of species, in 1775 he argued that all varieties of mankind belonged to a single species because of their mutual ability to reproduce, and he placed them in basically the same four units that had been recognized by Linnaeus. According to Kant, blond northern Europeans were adapted to the cold and damp. Americans were copper-colored and adapted to dry cold. Africans were black and adapted to dry heat, while Indians were olive-colored and also adapted to dry heat. Kant noted, accurately enough, that you didn't always find the same variety in the same environmental conditions; but he derived from that a conclusion exactly opposite to Buffon's: that, once particular characteristics were in place, there was no going back. Only the original "stem" form could be modified. This was the conclusion of a rationalist philosopher, not of a biologist; but Kant's essentialist views on race proved a burden that biologists would have difficulty throwing off.

One of Buffon's successors in the French zoological establishment was Jean-Baptiste Pierre Antoine de Monet, Chevalier de Lamarck, the first scientist to suggest that lineages of organisms had actually changed over time (albeit separately in different lineages, via an inner impulse). In 1809 Lamarck pro-posed that bipedal humans were descended from an apelike ancestor; and although he was not explicit about how modern human variety had come about, it was pretty clear he thought it had been through environmental accom-modation. Unlike his colleague and rival Georges Cuvier, who recognized three human "varieties" (white Caucasians, yellow Mongolians, and black Ethiopians), Lamarck recognized six: Caucasian (European), Hyperborean (the circumpolar peoples), Mongolian (Central Asians), American (Amerindian), Malayan (South Asian), and Ethiopian (African).

With his own addition of the Hyperborean group, Lamarck had almost cer-tainly based those categories on the work of the German physician Johannes Friedrich Blumenbach, a scientist who devoted the greater part of his career (spent almost entirely within the small city of Göttingen) to understanding the variety of humankind around the world. Blumenbach's main object of study was the human skull, of which he avidly collected examples from all over the planet; and his 1776 thesis, translated into English under the title *On the Natural Varieties of Humankind*, rapidly became the definitive exposition of the subject. In tandem with his belief that species could diversify under external influences, Blumenbach entirely concurred with Buffon's view of the unitary origin of humankind. As pointed out by the anthropologist Earl W. Count, it is thus unfortunate that his major conclusion was rendered into English as "the causes of degeneration are sufficient to explain the corporeal diversity of mankind," because the German word "Ausartung" that was trans-lated into English as "degeneration" (with all that term's unhelpful and unin-tended overtones of deterioration) might equally well have been translated as "divergence."

In the first edition of *Natural Varieties*, Blumenbach went with the four separ-ate kinds of modern human that Linnaeus had recognized, each one associ-ated with a particular geographic region. But by the time of the third edition, published in 1795, he had added a fifth. Prime among these five were the white, rosy-cheeked Caucasians, so called because Blumenbach believed that the origin of his European variety lay in the Caucasus, between the Black and Caspian Seas. Yellow-skinned and straight-haired, the Mongolians were the Asian peoples exclusive of the Arctic, India, and the islands of the Malay Archipelago. All of the black-skinned, curly haired sub-Saharan Africans belonged to the Ethiopian variety. The copper-colored and straight-haired American group embraced all of the peoples of the New World except the circumpolar populations, while the brown-skinned inhabitants of what are now Indonesia and the Pacific islands belonged to the Malayan group.

Importantly, this division implied no judgment or ranking of the units involved; and indeed, Blumenbach went out of his way to emphasize that Ethiopians were in no way inferior to the others "concerning healthy faculties of understanding, excellent natural talents and mental capacities." The "degeneration" had taken place from an ancestral Caucasian type, via accommodation to local conditions in different parts of the world: through the Malayan toward the Ethiopian in one branch, and through the American toward the Mongolian in the other. Echoing Buffon, Blumenbach also noted that "individual Africans differ as much, or even more, from other Africans as from Europeans." What is more, while Blumenbach was prepared to formally recognize and name his five varieties, he also perceived numerous "insensible transitions" among them. He clearly saw that boundaries within humankind were blurry, something that signaled to him more than ever that all humans belonged to the same species.

By the end of the eighteenth century, then, natural historians and philosophers had reached a pretty firm agreement on the basic unity of humankind. Certainly, various geographic varieties were distinguishable within the human species, most likely due to local environmental influence; and those varieties might be associated in some way with different temperaments or forms of social organization. Nonetheless, it was generally felt unwise to draw firm lines among them, or to rank them in terms of moral, physical, or intellectual superiority. This was a remarkably balanced view, especially given how little was objectively known and the hovering intellectual legacy of the Great Chain of Being. And then, politics intruded.

The Early Nineteenth Century

As the trickle of information from far-flung parts of the world became a flood, and as trading expeditions turned into extensive economic and political engagements, it was inevitable that the interpretation of human variety would change from a matter of intellectual interest to one deeply colored by practical considerations. This was particularly true in the fledgling United States, which had broken its political ties to Britain late in the eighteenth century but that, with the ongoing oppression of its native peoples, was still an outpost of European ethnicity. The economy of the United States was hugely dependent on the institution of slavery, mainly in the cotton plantations of the South. The slaves were people of African descent, many of whom had come via the sugar plantations of the Caribbean that the labor and sufferings of their forebears had made possible. In the land where all men were allegedly created equal, some justification was needed for a practice that flew so flagrantly in the face of human equality, and justification of this kind was often sought in claims for the "inferiority" of the people subjugated in this way. The idea that such inferiority was inherent, rather than acquired, was reinforced by the mounting practical experience of Europeans in colonial outposts, who often died instead of adapting to local environmental circumstances. To a resurgent generation of polygenists, the corresponding inference that differences between human geographic variants thus could not be accounted for by climatic accommodation suggested only one thing: that those variants were different species, or at least so deeply differentiated as to imply separate origins.

A leading proponent of this kind of thought was Samuel Morton, a Philadelphia anatomist and physician. Around 1820, Morton had begun to assemble a vast collection of human skulls from around the world; and in 1839, in a work titled *Crania Americana*, he presented his conclusions from them. Morton was puzzled by the fact that images of apparently fully differentiated black and white races existed in Egyptian tombs dating from 5,000 years ago, whereas, as calculated by the Irish Archbishop James Ussher in 1650, the world had been created only in 4004 BC. A thousand years, Morton reasoned, was scant time for the modern races to have differentiated, by Blumenbach's "degeneration" or by any other known mechanism. Better, surely, to conclude that the Creator had produced them all at once, each one for the particular environmental and "moral" circumstances in which it was going to live. By implication, Africans were born to be enslaved.

One question Morton assiduously avoided was whether the different kinds of humans were separate species, or merely "races," his term of choice. This was convenient, because he never quite sorted out what exactly races were, beyond being some kind of "primal variety." Others made a distinction between "races" and "varieties," the former seen as permanent, whereas the latter represented a more ephemeral and potentially reversible kind of division within the larger species unit. Varieties were thus becoming more like the subspecies we recognize today, whereas races were essentialist, there for the long haul.

Morton himself basically adopted Blumenbach's five-variety scheme, although he subdivided the five divisions into 22 "families." And his great departure from the spirit of Blumenbach's approach, taking the subdivision of humanity into an altogether different sphere, was to rank those varieties in the manner of the Great Chain of Being. According to Morton, the Africans were the "lowest grade of humanity," while Caucasians (including himself), "with the highest intellectual endowments," lay at the top of the series. The measurements (principally of brain size) on which this ranking was allegedly based have recently been the subject of controversy, and they were probably biased unconsciously – but the bias was certainly there.

Morton himself did not opine on the matter of slavery. But following his death in 1851, a decade before the Civil War started, his ideas were eagerly adopted by apologists for slavery. In 1854 his disciples Josiah Nott (a slave-owning surgeon) and George Gliddon (an avocational Egyptologist) published their broadside *Types of Mankind*, in which they argued specifically for separate creation, and for the biological superiority of whites over the blacks they enslaved. This probably did not mean too much in practical terms to those already lucratively involved in a deeply entrenched economic system that brought profit not only to the southern states but to communities all over the country (the northern and southern economies were intricately intertwined); but it would have provided some comfort to anyone prone to feelings of guilt. Even if Americans were not actively seeking moral justification for an economic system that brought them extraordinary wealth, justification of the kind Nott and Gliddon offered would have been welcome.

Intellectual support for the polygenic notion during the mid-nineteenth century also came from one of the United States' most distinguished scientists, the geologist Louis Agassiz, a Harvard professor of Swiss origin who had been one of the first to recognize the geological evidence for ancient Ice Ages in Europe.

It had apparently been a visceral experience for him to encounter black American domestic servants following his upbringing in homogeneous Switzerland, and he came to believe that the different groups of humankind must have been created separately. Accordingly, his opposition to slavery came not from any moral objection, but from his belief that it brought blacks and whites too close together.

Racism of this general kind was, of course, nothing new. In mid-1850s France, for example, the aristocratic Arthur de Gobineau published his *Essay on the Inequality of Human Races* in which he claimed that each of the three races ("black," "white," and "yellow") that he recognized was inseparable from its culture. The white (Aryan) race was superior to the others, and mixture between the races only led to biological and cultural deterioration. Looking around, Gobineau saw this process already in an advanced phase, and as the cause of myriad social problems. In Germany, Gobineau's ideas were to be seized upon by the Nazis in the next century, although they had to quote selectively from his writings because, fairly unusually for his time and place, he was not anti-Semitic. Across the English Channel in England, the superiority of Europeans was usually assumed rather than argued for; and most people were prepared to accept the biblical monogenist position, even as the nation increasingly profited from slavery overseas. Nobody really wanted to rock the intellectual boat, while only a minority was interesting in killing the goose that was reliably laying golden eggs. Early nineteenth-century Scotland was, in contrast, more freethinking. In Edinburgh, for example, Lamarck's ideas of change began early on to influence thinkers such as the anatomist Robert Grant, with whom the young Charles Darwin studied briefly in the 1820s.

3 Race after Darwin

Separate Origins

As the nineteenth century dawned, the idea that species might not forever remain as the Creator had made them was no longer unthinkable. At one end of the range of possibilities was Buffon's limited notion of within-species change; at the other was Lamarck's vision of lineages transforming themselves through inner impulse. What's more, Charles Darwin's own grandfather had even proposed, back in the 1790s, that "the final course of [the] contest among males seems to be, that the strongest and most active animal should propagate the species which should thus be improved." At this point, it had become more-or-less inevitable that someone would eventually come up with the evolutionary ideas that the younger Darwin expressed in *On the Origin of Species*, and that we summarized in Chapter 1. And it was even more inevitable that, once the evolutionary cat was out of the bag, it should have been seized on by apologists for slavery at a time when tensions surrounding the institution were approaching a peak in the United States. As the American Civil War raged, the zoologist Carl Vogt (who had studied with Louis Agassiz) published in London a work titled *Lectures on Man*, in which he vigorously promoted the belief of the German anatomist Hermann Klaatsch that the different races of humankind were descended from different species of ape. This notion naturally appealed to supporters of slavery such as the physician James Hunt and the anti-suffragist James McGrigor Allan, authors respectively of *The Negro's Place in Nature* and *Europeans, and Their Descendants in North America*. In turn, those popular works were eagerly seized on by others who supported the enslavement of blacks.

Inevitably, during the mid-nineteenth century the debate between American monogenists and polygenists was going to lose any hope of theological or scientific detachment. After all, *On the Origin of Species* was published just eight days after the leading abolitionist John Brown had been hanged in Charles Town, Virginia, for fomenting slave rebellion and raiding Harper's Ferry. The war between the northern and southern states was not fought explicitly on the issue of slavery, but in retrospect it symbolizes a struggle between two socioeconomic systems, one of which directly depended on slavery for the wealth it generated. Following the war, slavery was formally laid to rest, at least in the United States. But its end had done nothing to snuff out the social attitudes that slavery had engendered, or indeed the still-smoldering sense of deprivation and injustice that came with losing it. Over several centuries of slavery, two dismal but common tendencies among humans – to hate what you abuse and to blame the victim – had coalesced in southern whites to justify the horribly intimate kind of violence that was needed to keep the institution functioning. And the deeply embedded social attitudes thus engendered were not about to go away simply because their original cause had been forcibly abolished. As we are about to see, the events of the Reconstruction period that followed the American Civil War thus turned out to be just as much a stain on American history as slavery itself had been, adding a new level of hypocrisy to the mix.

As the southern states began to re-enter the Union between 1866 and 1870, the Thirteenth, Fourteenth, and Fifteenth Amendments to the Constitution abolished slavery, guaranteed equal rights to all citizens, and gave all males everywhere the vote. Enterprising former slaves accordingly started acquiring land and participating in economic and political life; and in 1872 the state of Louisiana elected its first (and still only) black governor. Yet, as a result of a horrendous political compromise, southern legislatures were soon passing the "Black Codes" that would curtail the newly won freedoms of the former slaves, in anticipation of the even more draconian "Jim Crow" laws that were to follow. More informally but equally menacingly, white supremacist groups such as the Ku Klux Klan were formed to terrorize the black population by means of beatings, lynchings, and the destruction of property, galvanizing a mass migration north.

Perhaps the most insidious and durable of all such developments involved institutions that had been originally established to recapture runaway slaves. Many modern American police departments are the direct or indirect descendants of organizations originally established for this purpose; and, consciously or otherwise, black Americans continue to this day to be disproportionately targeted by them. A "black" individual is three times more likely than a "white" one to be killed by a policeman. Blacks make up only 13 percent of the general population, but they account for one-third of those in prison and for fully 42 percent of Death Row inmates. Most white kids grow up thinking of the police as their protectors; black fathers find it necessary to warn their sons to give them a wide berth. While this book was being written, a violent mob of white insurrectionists stormed the Capitol Building in Washington, DC, the architectural embodiment of American democracy. At some entrances to the building, uniformed police reportedly stood politely aside to let Confederate flag-waving louts in to trash the interior, so that even well before the most sinister implications of this invasion had become evident, many African Americans were already asking themselves what kind of mayhem would have transpired if the looters had looked like them. Nobody was forgetting that policing in much of the United States had its origins in the nineteenth-century "slave patrols." A dreadful history remains dreadfully alive.

Huxley, Wallace, and Darwin

In 1863 the English comparative anatomist Thomas Henry Huxley, an enthusiastic evolutionist, became the first scientist to undertake a review of the handful of fossils that then comprised the human fossil record. The most important fossil available for comparison to modern human skulls was the Neanderthal skullcap that had been discovered in Germany in 1856. This skullcap had held a large brain, but in contrast to our own tall, rounded crania, it was long and low, with jutting brow ridges at the front and a distinct protrusion at the back. Hominins of this kind are nowadays assigned to their own species, *Homo neanderthalensis*. Yet Huxley was able to conclude that, "although the most pithecoid of human skulls, the Neanderthal cranium is by no means as isolated as it seems at first, but forms, in reality, the extreme term of a series leading gradually from it to the highest and best developed of human crania." In other words, the Great Chain of Being was back, with a vengeance. The races of humankind were related, but they could be rated as "lower" or "higher"; and the strange Neanderthaler, by Huxley's reckoning a member of a "barbarous race" that had inhabited Europe in ancient times, ranked at the bottom, just below a modern Australian aborigine. In reality, the anatomical gap between the Neanderthal and the Australian vastly exceeded that between the latter and a modern human from anywhere else. But Victorian England smugly ruled much of the world, and the paternalistic illusion of intrinsic superiority that went along with that political domination was evidently strong enough to influence the conclusions of a serious scientist.

Still, this episode shows equally that Huxley saw the human species as a single, if variable, entity. As did Alfred Russel Wallace, Darwin's co-originator of the concept of natural selection, who broached the issue in 1864. The widely traveled Wallace concluded that the differences between the races of humankind were of long standing, but that all traced back to an ancient "single homogeneous race" that was united by the common escape from the "great laws which irresistibly modify all other organic beings" conferred by our common possession of material culture.

Though appalled by the claims of Vogt and his like, the retiring Charles Darwin, scion of a stalwartly abolitionist family (Figure 3.1) was initially reluctant to address humankind specifically. But he eventually joined the fray in 1871, with his two-volume *Descent of Man*. This work was essentially an impassioned plea for the monogenist cause, by one who was a fervent opponent of slavery both by heritage and by inclination. "All the races," Darwin wrote, "agree in so many unimportant details of structure and in so many mental peculiarities that these can be accounted for only through inheritance from a common progenitor." Yet he, too, was a product of his times, happy to go along with the Victorian stereotyping of foreigners of all kinds, including the "squalid, unaspiring" Irish. Still, perhaps the most interesting aspect of this work is Darwin's failure to find that natural selection was an adequate explanation for the geographical variety of humankind. Climate, which had traditionally been adduced to explain them, was simply inadequate to explain the differences observed. This might have been one reason why he devoted much of his energy in this work to the issue of "sexual selection": change through mate choice.

Figure 3.1   Charles Darwin came from a committedly abolitionist family. This porcelain cameo, "Am I Not a Man and a Brother?" was first produced in the late eighteenth century by Darwin's father-in-law, the pottery manufacturer Josiah Wedgwood, and has been reissued many times since.

The Late Nineteenth and Early Twentieth Centuries

The most influential evolutionist in late nineteenth-century Germany was Ernst Haeckel, polymath and devotee of Darwin and natural selection. But although Haeckel correspondingly thought that all the varieties of humankind were descended ultimately from a single apelike ancestor, he also saw them as highly distinct. Each was descended separately, in a separate part of the world, from an ancestor that was still an ape. This meant that language, in his view the "special and principal" human adaptation, must have evolved separately in each lineage; but it had to have been this way, because Haeckel believed deeply in racial inequality. Borrowing from Gobineau, he identified the superior Aryan race with the tall, blond stereotype of the German *Volk*, destined to triumph in any struggle with inferior types. Haeckel was adamantly opposed by his equally eminent former teacher Rudolf Virchow, who elegantly demonstrated that the German people didn't actually fit the stereotype; but after Virchow's death in 1902, Haeckel had the field pretty much to himself. In 1904 he founded the anti-clerical Monist League that vigorously preached Aryan superiority, and the necessity to forestall its dilution. This message took on additional force when the draconian treaty terms imposed after World War I impoverished the German population, fostering widespread distress and feelings of grievance. Small wonder that the politicians would subsequently turn Haeckel's declaration that "history should be biology writ large," into prophesy.

In the English-speaking world, ideas derived from evolution developed in what was in some respects an equally disturbing fashion. But with a difference. In England, the emphasis remained mainly on the individual, where Darwin had placed it, rather than on the group. To Darwin's own dismay, a school of "social Darwinists" claimed that social policies aimed at helping the poor and the infirm only weakened society by permitting undesirable individuals to proliferate. It would be much better for society, and for the species in general, it was argued, if the authorities would just stand aside and let Nature take its course. From this point, it was not much of a leap to the advocacy of eugenics, a movement named in 1883 by Darwin's cousin Francis Galton. Eugenics sought to improve the species by promoting the breeding of individuals with desirable traits, giving rise in the process to its nastier counterpart, dysgenics, which sought to impede the breeding of the feeble of mind and body. All of this fit in well with the progressivist Victorian ethos, though Galton himself, disturbed by the "Nature red in tooth and claw" approach of the most enthusiastic social Darwinists, wrote that "I conceive it to fall well within [Man's] province to replace Natural Selection by other processes that are more merciful and not less effective." Nonetheless, the cat was already out of the bag, and although Galton had idealistically wished simply to foster excellence in society and the human species, dysgenics in the end had a far bigger impact than eugenics (really only ever tried out in practice by the Nazis) ever did. Well before the nineteenth century ended, governments on both sides of the Atlantic were forcibly sterilizing those considered undesirable – a practice that only began to be ended by post-World-War II revulsion with the Nazi atrocities.

An exception to the focus on the individual was Francis Galton's 1869 attempt to quantify the ability of different races to produce outstanding individuals. Among modern peoples the English pretty predictably came out on top in this analysis, with Africans lower down and indigenous Australians at the bottom. Awkwardly, though, the ancient Athenian Greeks outscored everybody, which said something fairly unflattering about the state of the modern world. This only served to inflate the worries of those in the United States who were concerned about the effects of immigration on the existing population. The immigrants were principally from Europe, and in his 1916 bestseller, *The Passing of the Great Race*, the patrician Madison Grant accordingly divided the Caucasian race into three varieties: Nordic, Alpine, and Mediterranean. Prime among them were the overachieving Nordics (Teutonic by origin), who had acquired their admirable features in bracing northern environments. The goal should accordingly be to augment this population and to prevent its adulteration in America by Alpines and Mediterraneans; and such was the influence of Grant's book that ultimately some 65,000 Americans were forcibly sterilized.

Vigorous opposition to Grant and his like came from the anthropologist Franz Boas. A German of Jewish descent who had studied with Virchow, Boas was a major influence in American anthropology in the decades around the turn of the twentieth century. He saw culture as a much more important influence than biology in shaping the human species; and soon after Adolf Hitler had assumed Germany's Chancellorship in 1933 he had begun to worry about how Jews would fare in the Nazi state. Those fears were realized almost immediately, as the purging began. In 1935 the Nuremberg Laws made marriage possible only between those who had been racially and physically certified, complementing the 1933 Law for the Prevention of Diseased Offspring. And by 1939 the carnage was fully underway, the killing of mentally and physically handicapped children soon extending to adult psychiatric patients and then to Jews, homosexuals, Gypsies, and anyone else the Nazis judged unacceptable. Many British academics sprang to support Boas as he condemned such horrors, but in prewar America he found a curious lack of concern among Gentile colleagues. In the end, only the Harvard anthropologist Earnest Hooton was prepared to stand publicly with him: maybe oddly, because Hooton also served on the notorious Committee on the Negro. An offshoot of the American Association of Physical Anthropologists (AAPA) and the National Research Council, this body concluded in 1937, after 11 years of deliberation, that "the negro race is phylogenetically a closer approach to primitive man than the white race." Some took exception, and at its 1939 annual meeting the AAPA considered a complex motion to reject a link between human physical and cultural differences, to repudiate the Aryan and Semitic racial categories, and to deplore racism. The motion was never passed.

One effect of the Nazi purges was to drive the secular Jewish anatomist Franz Weidenreich from the University of Heidelberg to Chicago, and ultimately to China, where he arrived in 1936. He was there to teach anatomy and to run the Cenozoic Research Laboratory in Peking (Beijing), the center for the study of the newly discovered Peking Man *Homo erectus* fossils. During his five years in Beijing Weidenreich developed a "polycentric" view of human diversification, whereby five geographic races of *Homo sapiens* (Australian, African, San "Bushmen," Asians, and Europeans) had evolved regionally and in parallel from an already variable *Homo erectus* ancestor. The various races thus had very deep histories, but nonetheless all belonged to *Homo sapiens*. This is problematic, because it is hard to imagine how different lineages of the same species could contrive to cross the speciation barrier, only to end up back in the same species. And Weidenreich's late summary of this process is consequently a bit opaque: "an already world-wide distribution of early phases . . . transmuted into more advanced types by vertical differentiation, while they split into geographical groups by horizontal differentiation . . . the main racial groups of today developed in parallel lines from more primitive human forms." Still, despite its questionable aspects this sibylline statement was later developed into the "single-species hypothesis" that held that only one kind of hominin could exist at any one time, and ultimately morphed into the "multiregional evolution" refinement which was to have some currency during the 1970s and 1980s.

To be fair we should note that Weidenreich was a human anatomist, and thus not particularly interested in matters of biological diversity; and maybe it is significant that his conclusion that "the facts available do not indicate that environmental conditions played a decisive role [in human evolution]" was approved mainly by his fellow anatomist Sir Arthur Keith. Otherwise, the two anatomists made rather odd bedfellows, and Keith went substantially further. Just as Weidenreich was beginning his work in China, Keith wrote that "throughout the Pleistocene period the separated branches of the human family appear to have been unfolding a programme of latent qualities inherited from a common ancestor of an earlier period." In other words, the geographical varieties of mankind had extraordinarily deep roots, and their fates were governed by biology. Maybe it's hardly surprising that, after the conclusion of World War II, Keith became a minor apologist for the "patriots" who had been condemned at Nuremberg.

Postwar Developments

As a result of the appalling excesses of the Nazis, both "race science" and eugenics fell from favor postwar. More in tune with the times was *We Europeans: A Survey of "Racial" Problems*, a book that the biologist Julian Huxley and the anthropologist Alfred Haddon had published in London back in 1935. The two scientists vigorously attacked the race science advocated by Madison Grant in the United States, and by the National Socialists in Germany. They pointed out that "pure races" were entirely hypothetical concepts, with no existence in the real world, and they suggested that any attempts to classify human variety would inevitably stumble on the issue of continuous variation. They readily admitted that "races" existed in the very broad sense that people with roots in different parts of the world looked rather different; but only in that sense: When you looked more closely, the lines among them blurred and disappeared, and you could find almost any "racial" character in combination with almost any other.

An early supporter of this important and ultimately influential viewpoint was Ashley Montagu, who had studied with Boas and who published his most famous work, *Man's Most Dangerous Myth: The Fallacy of Race*, as early as 1942. After the war, Montagu was hired by UNESCO to coordinate an international panel to prepare a statement on race to be approved by the Assembly of the United Nations. Consensus turned out to be surprisingly hard to find, and only in 1950 did UNESCO come out with a pretty anodyne declaration that all humans belonged to the single species *Homo sapiens*, and that this species contained genetically and culturally distinctive populations that also varied internally. The report also tried to shift the issue from the biological to the sociocultural sphere by suggesting that the term "race" be replaced by the less loaded term "ethnic group," emphasizing that "the biological differences between ethnic groups should be disregarded from the standpoint of social acceptance and social action ... [T]here is no proof that the groups of mankind differ in their innate mental characteristics." An uproar nonetheless ensued, and another committee (this time not carefully curated by Montagu) was convened to produce a revised statement. Consensus was elusive, and an early draft was abruptly withdrawn. When the final report appeared in 1951, it declared that where human populations showed "well-developed developed and primarily heritable physical differences" from others they could indeed be called "races," remaining vague on such matters as the roles of "innate capacity" versus "environmental opportunity" in IQ test performance.

The difficulties experienced by UNESCO in finding a consensus on the matter of race mirrored a tension that was developing between the anatomists who had dominated prewar biological anthropology and the postwar practitioners of the "New Physical Anthropology" who favored a more dynamic view of human populations. The English biologist J. S. Weiner put the new view nicely in 1957, when he characterized *Homo sapiens* as "a widespread network of more-or-less interrelated, ecologically adapted, and functional entities." In America a leader of the new movement was Sherwood Washburn, who had fallen under the influence of Theodosius Dobzhansky, one of the founders of the Modern Evolutionary Synthesis (see Chapter 1), when he took up a post in the medical school at Columbia University where Dobzhansky had his lab. Finally, a portal had opened between evolutionary theory and physical anthropology. As a geneticist, Dobzhansky himself was more interested in the distribution of characters such as blood groups among the populations that made up humankind than in the "races" that he defined as "group[s] of individuals which inhabit a certain territory and which [are] genetically different from other geographically limited groups." In his *Mankind Evolving*, published in 1962, Dobzhansky accepted human "race differences" as "facts of nature," pointing out that the human species is far from unique in being "polytypic" – that is, consisting of several physically and genetically distinguishable local variants (whether "races" or "subspecies," Dobzhansky didn't care). Some of the distinguishing features might have been due to selection, others to drift. But because everyone belongs to the same interbreeding species, no race could be seen as "pure." And all were transient in evolutionary time.

Some of those who wondered why humans vary from place to place nonetheless continued to feel that classification might provide a useful heuristic. The developmental anthropologist Stanley Garn, for one, concluded that it might be helpful to distinguish between large "geographic races," smaller "local races," and tiny "micro-races" in the attempt to understand what underlay the physical differences observed among populations. He made a heroic effort to do so, but soon got bogged down in a morass of splintering categories that made it pretty clear what a hopeless quest he had embarked on. At the other end of the spectrum was the genetic anthropologist Frank Livingstone, author of an influential 1962 article bluntly titled *On the Non-Existence of Human Races*, which argued not only for entirely abandoning the race concept but for replacing it by the study of how individual characters varied geographically. As Livingstone neatly put it: "There are no races, only clines" (clines being geographic gradients in gene frequencies).

But the categorizing just wouldn't go away. In 1964 the Philadelphia anthropologist Carleton Coon published a fat tome called *The Origin of Races* in which he exhaustively analyzed the entire human fossil record to demonstrate that Weidenreich had been right to conclude that the origins of the five modern human races had been very deep, back in the time of *Homo erectus*. To accommodate the Synthesis, Coon imagined a scenario in which adaptation to local circumstances (hence the differences) had been complemented by "genetic contact [among] sister populations" (hence the unitary species). Because of his model's emphasis on continuity, Coon was obliged to find a criterion to distinguish between *H. erectus* and *H. sapiens*, and he settled on brain size, placing the threshold at 1,250 mL (the modern average is about 1,330 mL, whereas that of the *H. erectus* crania he had to hand was around

1,000 ml). Looking at the available evidence, Coon discovered that the different races had crossed the *H. sapiens* threshold at different times: Asians first and Africans last. Unsurprisingly, few liked any of this. Coon was widely excoriated both for racism and for poor scholarship; but in the next year, undaunted, he pursued his theories even further in *The Living Races of Man*. The timing of this volume could hardly have been worse: 1965 was, after all, the year of the violent events of Bloody Sunday, in which peaceful marchers for black civil rights were mercilessly beaten by troopers as they tried to cross the Edmund Pettus Bridge outside Selma, Alabama. Small wonder that the debate over Coon's claims became so acrimonious that many human biologists began shunning the study of race in favor of the less fraught area of human adaptation.

One of Coon's leading critics was the anthropologist Loring Brace, who avidly advocated the "single-species hypothesis" from the mid-1960s on. Dobzhansky had claimed back in the 1940s that the human ecological niche was so broad as to obviate the presence on the ecological stage of more than one species at a time, and Brace energetically preached that all known human fossils belonged in one single lineage, a claim that only had the shakiest empirical support. Nonetheless, the cry was soon taken up by Brace's University of Michigan colleague Milford Wolpoff, who was subsequently obliged to engineer the fallback hypothesis known as "multiregional evolution." Born of a generous ecumenical feeling that compelled its proponents to consciously disassociate themselves from Coon, the multiregional notion harked directly back to Weidenreich in viewing the geographical variants of mankind as being of ancient derivation. Each geographical race had evolved *in situ* from a *Homo erectus* ancestor in response to local exigencies, while still interbreeding enough at the population edges to maintain genetic continuity. Eventually, the multiregionalists sidestepped the species boundary problem by the simple expedient of subsuming the ancient and distinctive *Homo erectus* into *H. sapiens*. In retrospect, this extraordinary gambit was clearly doomed from the start; but it nonetheless had the effect of delaying the general acceptance of what we now recognize: that *Homo sapiens* is an extremely young species that emerged in Africa some 200,000 years ago, and that almost all human physical variation, worldwide, is an epiphenomenon of the last 100,000 years.

4 Race in the Era of Genetics and Genomics

Early Genetics

As scientific knowledge increased, the outward physical variation of humans on this planet posed a mystery to natural historians, and begged explanation. Why were there so many different "kinds" of people on the planet? If scientists could understand how this variation worked in nature, then perhaps, the reasoning went, we could understand *why* people appear different. Implicit in understanding why people differ across the world was the prospect that this understanding would allow better categorization of humans, just as it was acknowledged to be important to categorize the variety of animals in nature.

While anthropologists focused on anatomical and behavioral traits, a parallel approach was developed by the first geneticists. Prior to the work of Gregor Mendel (Chapter 1) there was no formal scientific basis for understanding how organisms inherited their traits. Most scientists interested in human variation before the rediscovery of Mendel's laws had been drawn to human disorders, malformations, and maladies that appeared to be passed on generation to generation. One of the first traits to be examined for its persistence in families was polydactyly, the occurrence of supernumerary fingers and toes. Examination of this malformation was followed by studies on "porcupine men" (an affliction known as *ichthyosis hystrix gravior*), skin disorders, and neurological disorders such as Huntington's chorea, now Huntington's disease. The maladies experienced by the inbred royal families of the time also attracted the attention of scientists interested in explanations for the patterns and distributions of medically important traits. We can envisage developments in the field as the opening of a succession of doors that led to the "racial science" that persists today in transmogrified form.

Door to "Racial Science" Number One: Galton and Eugenics

In 1814, half a century before Mendel, Joseph Adams, a British physician, attempted to categorize disorders that he felt were hereditary (meaning passed on from one generation to another). Adams's scheme was quite prescient, especially in the absence of a known mechanism for heredity. He categorized heritable disorders separately from familial disorders. The former appeared in multiple generations of a family tree, while the latter occurred in siblings of a single generation. Adams also distinguished between disposition (the propensity to develop a disorder during an individual lifespan) and whether a disorder was congenital (present at birth). Charles Darwin later considered hereditary disorders and their modes of inheritance in his book *The Variation of Animals and Plants Under Domestication*, classifying at least eight human disorders as hereditary. Darwin's mathematically inclined cousin and contemporary, Francis Galton, developed ideas about particulate inheritance (sometimes right and sometimes wrong) that were accurate enough for him to be widely regarded today as the father of the field of quantitative genetics (and, unfortunately, of eugenics too). Galton was explicit, though, that the goal of his work was to "know thyself." Galton and his students, such as Karl Pearson, posited that it was possible to improve the species, and part of this endeavor involved the ability to categorize and systematize human variation. It was in the context of the betterment of the species that Galton opened the door to eugenics, a theme that would be repeatedly politicized over the next 140 years or so.

August Weismann, a German cell biologist interested in animal development, gave the concept of heredity a cellular component by proposing his "continuity of germ plasm" concept in 1885. This concept posited the existence of two kinds of cells in organisms – soma and germ – and that only the germ plasm was relevant to inheritance. At around the same time, Mendel was toiling away in his monastery garden, taking copious notes and manipulating data from crosses on pea plants. But while the work of Darwin, Galton, and Weismann saw the light of day immediately, the principles of Mendelian genetics did not emerge until the turn of the twentieth century. All of these developments are crucial to the genetic understanding of human biology that in turn is critical to our understanding of biological race.

Door to "Racial Science" Number Two: Garrod and Inborn Errors of Metabolism

Scientists interested in human variation were finally introduced to a well-defined and scientifically verified mechanism by the rediscovery of Mendel's laws of inheritance at the beginning of the twentieth century. More refined tools for studying heritable variation were subsequently developed through advances in cell and developmental biology. Throughout, human disorders remained the focus. Not only were outward physical disorders like Duchenne muscular dystrophy and albinism studied, but inborn metabolic disorders were also examined. Archibald Edward Garrod, a British physician, was fascinated by the metabolic disorders some of his patients exhibited. In 1902 he published a paper entitled "The incidence of alkaptonuria: A study in chemical individuality," in which he recognized that disruptions of the chemical or biochemical aspects of human physiology were also part of the human variation spectrum. In that paper he asked, in the context of alkaptonuria (a disorder in which homogentisic acid accumulates in the body) and other metabolic disorders: "May it not well be that there are other such chemical abnormalities which are attended by no obvious peculiarities and which could be revealed by chemical analysis?" Here, Garrod opened the door for reductionism in the study of human population variation by suggesting that we can learn much about human variation (maybe even more than from outward physical appearance) from such abnormalities. This sentiment is still very much alive among geneticists and genomicists interested in human population variation. The reason reductionism is important in this context lies in its capacity to oversimplify rather complex genetic interactions involved in producing phenotypes of organisms, including those of humans. If a genetic problem can be reduced to a system with a simple explanation, then it becomes easier to quantitate and categorize complex traits like behaviors. Such categorization leads to arguments about inherent abilities in different groups of people and, further, to scaling of those differences as superior or inferior.

The ability to peer inside human cells using microscopes and specialized ways of staining cells that made the individual structures of the cells clearly visible led to the discovery of chromosomes. The concept that chromosomes carried heritable information was a very logical inference from these observations because some human disorders were correlated with changes in chromosome morphology in the cell. On the heels of the development of this approach to studying human inherited variation came the discovery of blood groups and the development of population genetics.

Door to "Racial Science" Number Three: Blood Groups

In 1900, the Viennese biologist Karl Landsteiner discovered the now well-known human ABO blood groups. Before his landmark work, medical scientists had tried to transfuse animal blood into humans, but with disastrous results because the blood cells clumped together immediately upon mixing. This clumping, known as "blood agglutination," also occurred in some cases of human-to-human transfusion, which prompted the notion that some humans had different kinds of blood than others, and that mixing them caused the clumping. Landsteiner identified the different blood types A, B, and O, now known to be the most clinically significant of many. These blood groups can be considered as alleles (alternative states of a particular gene) that produce six genotypes – AA, BB, OO, AB, AO, and BO. Two decades later, Ludwig and Hanka Hirschfeld published an analysis of the ABO blood group frequencies in different "races." Perhaps more telling than their results (that Indians, English, French, Italians, Germans, Austrians, Bulgarians, Serbians, Greeks, Arabs, Turks, Russians, Jews, Malagasy, Negroes, Annamese (Vietnamese), and Indians are differentiated from each other by blood group frequencies) was the difficulty in choosing appropriate populations to be analyzed. And it was at this point that the door was cracked open for the use of genetics in racial biology.

Door to "Racial Science" Number Four: Human Allozyme Variation

Three biologists, the American Sewall Wright and the English J. B. S. Haldane and R. A. Fisher, established the field of population genetics early in the twentieth century, basing it primarily on the Hardy–Weinberg Theorem. In

1939, Fisher and his colleague George Taylor used the results of blood group typing to make one of the first studies of population differences in England and Scotland, and events subsequently moved quickly. Such early work on blood groups quickly exposed all the many problems that we will encounter when we address the frequencies of DNA sequence polymorphisms in human populations. Those problems include the difficulty of defining specific populations, and of interpreting their gene frequencies in a global context. Nonetheless, researchers clung doggedly to the approach, and expanded its use to the reconstruction of the divergence and relationships of various groups of humans.

Biochemical genetic markers were almost completely limited to blood groups until the early 1960s; but that changed when Harry Harris, a British human geneticist, decided that he could do better. Harris knew a lot about human populations, and he also knew that the proteins that we produce for normal cellular function are encoded for by inherited genes. He also knew a bit about protein biochemistry, including that minor changes in the sequence of a gene might cause a slight change in the resulting protein, a change that would manifest itself as a change in the protein's physiochemical properties. Harris also knew that if you applied an electric field to such proteins in a gel matrix, the change in the amino acid sequence would alter the movement of the protein through the matrix, with some forms of the same protein moving farther than others. In other words, how far a protein moves in a gel is a function of the amino acid sequence present in the protein.

While there are only 35 different blood groups in human populations, tens of thousands of proteins are coded for by the human genome, which are involved in human physiology and development. By running protein extracts from people on those gel matrices, researchers can discern subtle differences in the proteins involved, and those differences can be correlated to the genes that they encode. By the 1970s, Harris's approach was called "allozyme analysis" (allozymes are different forms of the same protein at a locus), and it became the workhorse of genetic research in all kinds of populations, most famously human populations. In 1994 Luigi Luca Cavalli-Sforza, Paolo Menozzi, and Alberto Piazza published a fat tome entitled *The History and Geography of Human Genes*, which detailed how human populations might have diverged given the allozyme results and blood group results. Characterizing the patterns of evolution of these proteins and blood groups (many were important in human medical biology) would have been plenty, but they went an extra step and used the data to make inferences about the relationships among populations of people on the planet.

Door to "Racial Science" Number Five: The 1960s and 1970s

We will discuss the value and validity of inferring hierarchical relationships of human populations later in this chapter, but it was during the 1960s and 1970s that the door swung wide open for racial genetic work. Some researchers, such as Richard Lewontin, tried to push the door back closed. His classic and aptly titled 1974 book *The Genetic Basis of Evolutionary Change* was inspired by his own allozyme work (with his colleague Jack Hubby) on fruit flies; but the technique was also widely adopted by scientists interested in humans, and especially in understanding the role of rare alleles in human populations. Not content with the evolutionary history of his flies, Lewontin delved into the growing human allozyme literature and examined the data closely. In a 1972 paper entitled "The apportionment of human diversity," Lewontin concluded that there is more variation within predefined racial groups than between them, meaning that there is a high probability that any two randomly selected people in the same "racial" group will be more distantly related to each other than either is to an individual in another group. As we will see, this "more within than between" dictum violates any attempt to infer hierarchical relationships among populations of humans. Three decades later, the geneticist A. W. F. Edwards attempted to rebut this dictum, suggesting that even though Lewontin's observation might be true, there might be underlying correlations hidden in the data that might be useful in differentiating human populations from each other using genes. With this suggestion, the door to racial genetics using biochemical methods started swinging back and forth.

At about the same time the allozyme research was expanding, the systematics researchers who study diversity in the living world were developing rigorous tools for reconstructing the phylogenetic relationships of organisms. The focus of these methods was the production of phylogenetic trees. Such trees are branching diagrams that represent the patterns of evolutionary divergence of the organisms being examined. While phylogenetic trees had appeared in the literature well before this time, those trees had not been constructed using rigorous methodologies. The branching diagrams that were produced up to the expansion of phylogenetic methodology in the late 1960s and early 1970s were based almost entirely on a researcher's subjective knowledge of the organisms included in the diagram. What allowed rigor to be added to the business of making phylogenetic trees was the addition of objective criteria that allowed researchers to judge one possible tree to be "better" than another. Three major approaches were developed during this time to do this: (1) the use of similarity as a tool for grouping; (2) statistical likelihood; and (3) parsimony. All three approaches use comparative data to generate the results they express, and all three result in branching diagrams. Following are brief descriptions of the three approaches. Two of these approaches (likelihood and parsimony) use what is called character state data obtained from DNA or protein sequences or from morphological data. The third (similarity) can also use character state data from sequences or morphology, but the data need to be transformed into some measure of similarity before proceeding.

Similarity: Data are collected for the organisms under examination, and some measure of the similarity of the information between organisms in the study is determined. For instance, one can count the number of differences in a DNA sequence or a protein sequence among organisms, and use that to cluster the organisms in the study. Other approaches that have been used in human-oriented phylogenetic studies have been measure of immunological distance or DNA–DNA hybridization approaches. Neither of these give character state data, but rather due to the nature of data collection for these approaches the data represent pairwise raw similarities between taxa in a study. Two organisms with greater similarity will cluster together, and the summary diagram will show them in close proximity and even sometimes connected via a branch. The resulting similarity-based diagram is called a dendrogram. Similarity approaches led later to the application of other clustering techniques, such as principal component analysis (PCA), to molecular data, as well as the more recent STRUCTURE approach. These methods, known as data reduction approaches, simplify the information into visibly digestible patterns that are often regarded as depicting the reality of genetic and genomic variation. PCA determines which of the variables or data points explain the distribution of the data and then uses only those components of the overall data set. Again, these approaches reduce the complexity of the overall data set so that a visible pattern can be extracted from the data. We challenge this notion of clustering and structuring, and hence the relevance to racial clustering using this approach, in Chapter 6.

Statistical likelihood: In this approach, a probability is assigned to the likelihood of the protein or DNA sequence data, given a specific branching pattern and a model by which the data change. What this means is that each sliver of data is evaluated for the probability of its fit to a particular branching pattern. Branching patterns get more and more complex as the number of terminals or taxa in the tree increase (Figure 4.1). For instance, while there are only 3 tree

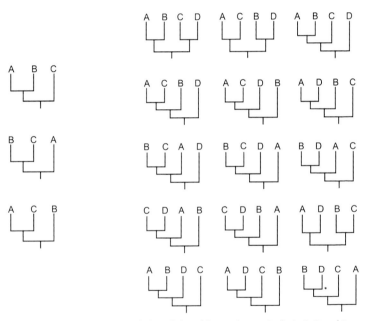

Figure 4.1 The possible trees with three (left) and four (right) terminals. A, B, C, and D are hypothetical biological entities.

topologies for 3 taxa, there are 15 trees for 4 taxa and 105 for 5 taxa. The number of trees to be evaluated increases roughly by an order of magnitude upon addition of another terminal. By the time there are 100 taxa in an analysis, there are more trees to evaluate than stars in the known universe. Computational tricks are used to solve this rather intractable problem. After all the slivers of the data set have been evaluated over the various tree topologies under consideration, an overall probability is computed. The result gives the branching pattern with the highest probability, and in such diagrams organisms that are connected to each other via branches are thought to be closely related.

Parsimony: As with likelihood, each sliver of sequence data or morphological data are evaluated for how well they fit each of the different tree topologies. The evaluation reveals the number of steps (changes) each sliver of data takes to find its place on each of the trees examined. As with the likelihood method, one wants to evaluate all possible permutations of branching between the organisms involved. By tracing where and how many times a trait needs to

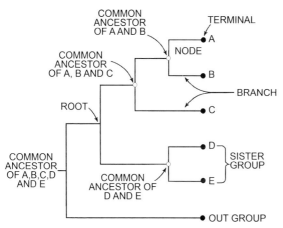

Figure 4.2   A hypothetical phylogenetic tree showing some of the common terminology used when viewing a tree.

change to accommodate the tree topology under consideration one can obtain an overall number of changes or steps that the character needs to fit the tree. When all of the traits are treated this way one can find the tree that requires the fewest steps, which is then judged the most parsimonious. The most parsimonious tree is consequently accepted as the best representation of the evolutionary history of the organisms being examined. For our purposes here, what the branching diagrams mean or tell us is more important than how they are constructed. Figure 4.2 shows a phylogenetic tree and some of the common terminology used when viewing a tree. We have taken this digression into systematics because many studies that use racial groups in genetic and genomic analysis fall back on the use of branching diagrams to summarize the relationships of the groups involved. We will return to the issue of trees in the context of race in the next chapter.

Door to "Racial Science" Number Six: Mitochondrial Eve and Y Chromosomal Adam

As the allozyme movement was becoming embedded in the regular science of evolutionary biology, and population genetics and systematics were formulating rigorous and repeatable methods to make inferences, some pioneering researchers in molecular biology were developing techniques to obtain the primary base sequences of genes for use in population genetics. Because determining the DNA sequences of organisms was very difficult in the 1970s and 1980s, many scientists adopted a shortcut method for DNA sequence analysis known as restriction fragment length polymorphism (RFLP). Restriction enzymes recognize specific DNA sequences and sever the DNA sequence molecule in or near the sites each one recognizes. For instance, the enzyme known as EcoRI cuts at the sequence GAATTC, and the restriction enzyme HindIII will cut at the sequence AAGCTT. Laboratory methods for detecting where these enzymes cut in DNA were developed and maps of where the various restriction sites exist were made. If an individual lacks the right sequence for cutting with the enzyme, then the maps will look different. Methods for characterizing polymorphisms using RFLPs were quickly developed, and it became relatively easy to assay specific genes for the presence or absence of restriction sites for large numbers of different restriction enzymes in large numbers of individuals.

Researchers discovered very early on that the DNA from the mitochondrion (mtDNA) could be much more easily manipulated in the lab than the DNA residing in the nucleus. Nearly every cell in the body has mitochondria, which are organelles lying in the cytoplasm outside the nucleus and which function as the powerhouses of cells. There are anywhere from 100 to 10,000 mitochondria in a cell, depending on the tissue you are looking at. Each mitochondrion has a small circular genome of its own. In humans this genome is about 16,000 base pairs long and codes for 13 proteins important in basic cellular respiration, and for several structural RNAs important in protein synthesis in the mitochondrion.

The mtDNA molecules have a peculiar pattern of inheritance. Unlike nuclear DNA, which comes from both parents, the mitochondrial genome is passed on from mother to offspring in a process called maternal inheritance. mtDNA is clonal, meaning that the genetic material is simply copied and passed along to the next generation, and the only variation that is introduced into this process over time is caused by mutation. In addition, its clonality and maternal inheritance means that the mitochondrial genome is a marker for female lineages, as males are evolutionary dead ends with respect to mtDNA as they do not transmit their mtDNA to their descendants.

Mutation is hardly uncommon, of course; but mtDNA has turned out to be something of a paradox of variation, as some parts of it evolve very rapidly (i.e., accumulate mutations at a high rate) and other parts of it evolve very slowly (incurring very few changes over time). One part of the mitochondrial genome that evolves very rapidly is one that does not code for a protein. This noncoding region evolves about twice as rapidly as the protein-coding genes in the mtDNA genome and has proven very useful for analysis at the population level. It is called the D-Loop, where D stands for displacement, because as the mitochondrial DNA replicates it displaces the two strands of the double-helix to start replication, making a loop visible. This region is also referred to as the control region, because it is the part of the genome that controls replication. Within the control region are highly variable stretches of DNA called hypervariable regions (HVRs); sometimes even closely related individuals in a population will differ in these HVRs.

In 1987 Rebecca Cann and Alan Wilson did the first detailed analysis of human mitochondrial DNA sequences using RFLPs. Cann and Wilson took tissue samples from 147 humans from diverse geographic locations and mapped hundreds of DNA polymorphisms in the mitochondrial genome (Figure 4.3). They used the presence or absence of 93 restriction sites as data to construct a branching diagram for human mtDNA. There are three major concepts to take home from this study.

First, trees should have a root. The Cann–Wilson tree is rooted between a group of seven people from Africa, and all the other individuals in the study. This root is contentious, but it is altogether reasonable given the way human populations moved around the globe: As we saw in Chapter 1, humans belong to a species that has African origins, and which subsequently migrated to all corners of the planet. The placing of the root of any tree is critical because it gives polarity or direction to the patterns of evolution inferred for the organisms being studied (Figure 4.4). Rooting a tree is at once simple (you should be able to root any tree with organisms that are not members of the group you are examining; these are called outgroups) and complex (if they are too far from the organisms of interest, the root will be essentially random).

Second, it appears that all 147 humans in the tree coalesce back to a single point or ancestor. Coalescing to a common ancestor is an important concept in population genetics and phylogenetics. It simply means that a group of descendants in an analysis can be traced back to a single ancestor. And, in fact, it is quite easy to reconstruct what that organism's mtDNA looked like. This coalescence prompted the authors to claim that "All these mitochondrial DNAs stem from one woman who is postulated to have lived about 200,000 years ago, probably in Africa" and it is the origin of the familiar term "mitochondrial Eve." Well, perhaps; but more likely not. When you are using a clonal marker like mtDNA, there is a high probability that it will coalesce to a single lineage rather than a single individual.

Third, note that no single geographic region is found in only one cluster of the tree. For instance, people from Africa, while exclusively showing up in that cluster at the root of the tree, are also found sprinkled throughout the rest of the tree. Systematists use the principle of monophyly to judge whether suspected groups of organisms actually demarcate "real" groups. Real groups to

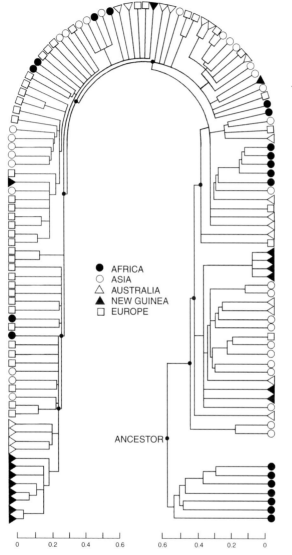

0   0.2   0.4   0.6   0.6   0.4   0.2   0

Figure 4.3 Phylogenetic tree of 147 human mtDNA restriction fragment polymorphism profiles.

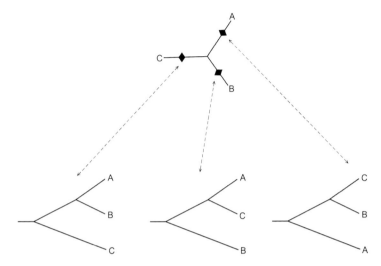

Figure 4.4   Rooting and polarity. A, B, and C refer to three hypothetical biological entities. The tree at the top is unrooted. The diamonds on the branches are the three possible rooting points in the unrooted tree. The resultant trees when rooted at the diamonds are shown below the unrooted tree. When rooting with C: A and B are sisters; when rooting with B: C and A are sisters; when rooting with A: B and C are sisters.

systematists are called "monophyletic," which simply means that all the organisms in a suspected group come from a single common ancestor. To apply this principle, a researcher needs to have a suspicion or hypothesis about how the organisms in a study are distributed. For instance, in the case of the Cann–Wilson tree we can use geography as a basis for formulating a hypothesis. Geography is often an improper guide for demarcating organisms because they can of course move about; but for this example it will suffice. One hypothesis, then, is that people from Africa are divergent, and distinct from people from other parts of the world. The test of that hypothesis is whether or not all people from Africa belong to a monophyletic group (i.e., come from a single common ancestor to the exclusion of all other people). And while some of the individuals from Africa do indeed reside in a distinct cluster at the root of the Cann–Wilson tree, there are several other people from Africa in the tree who destroy the notion of African monophyly, and the hypothesis can be rejected. Likewise, hypotheses about the monophyly of people from other geographic localities can also be easily rejected. It turns out that the business of using trees to decipher the biology of organisms within a species can be a very tricky endeavor indeed.

Since the initial "mitochondrial Eve" observation, hundreds of thousands of human mtDNA genes and genomes have been sequenced. Organizing this information into interpretable patterns has been a major preoccupation of human geneticists studying mtDNA, and it is directly relevant to our views of race. Because mtDNA is uniparentally inherited – that is, it is passed on from mother to offspring – it is called haploid, and any marker established using mtDNA is haploid in nature and is called a haplotype. The research here is inherently racial as haplotypes are often established, and interpreted, based on the geographic or assigned "racial" origin of the people analyzed. This major data problem was sidestepped by recognizing that a person's mtDNA (whether from RFLPs, D-Loop sequences, HRVs, or whole mtDNA genome sequences) can easily be categorized into what are called haplogroups. Haplogroups are so named because they are based on haploid (uniparental) mtDNA or Y chromosomal variation. As we will soon see, the human Y chromosome, which is passed on from father to son in a process called paternal inheritance, is also a source of haplogroups. Any human mtDNA or Y chromosomal DNA can be sorted into such haplogroups, opening the door for rapid characterization of a person's haplogroup membership.

Millions of mtDNA sequences exist in the main database for DNA sequence information at the National Center for Biotechnology Information, and other databases. One of these databases is a website dedicated to human mitochondrial DNA, called MITOMAP. Remember that Cann and Wilson used about 150 humans in their initial study; MITOMAP archives information for over 50,000 fully sequenced mitochondrial genomes. Y chromosomal variants are archived at the International Society of Genetic Genealogy website. As of the writing of this book there are 5,400 mtDNA haplogroups and 2,504 Y chromosomal haplogroups, with many haplotypes in each haplogroup. In

Chapter 5 we discuss in detail what we know about the thousands of both maternally inherited and paternally inherited haplogroups.

Door to "Racial Science" Number Seven: Human Variation Panels

Part of the reason why the more recent Y chromosomal and mtDNA studies have been so informative is the availability of standardized human tissue collections. These collections provide researchers with standardized panels of DNA data from diverse human populations that best serve the exploration of variation in our species. These collections, also called "human population-based biobanks," archive and store samples from multiple humans from multiple geographic locations. Well-archived human biological materials have become abundant since the time of the first organized attempts like the Framingham Heart Study, established in 1948 and maintained to the present.

Researchers in the 1980s and early 1990s recognized two reasons why standardized collections for use in characterizing human diversity are important. First, a standardized collection would provide researchers with the same samples for the many different approaches they were taking. Second, a standardized collection would make access to such samples quick, easy, and inexpensive (or at least less expensive than traveling to the far reaches of the planet to collect samples). In 1983, the Centre d'Etude du Polymorphisme Humain (CEPH; now known as the Fondation Jean Dausset-CEPH) was initiated by Jean Dausset and David Cohen. In 1991, Luca Cavalli-Sforza, along with Allan Wilson, hatched the idea of a Human Genetic Diversity project, or HGDP. In April 2002, the two collections collaborated to produce a panel of over 1,063 human samples from 52 or so researcher-defined geographic regions, which is described as the HGDP–CEPH panel. The samples were obtained with informed consent, and the privacy of the people who contributed them remains protected. The samples were brought into the lab, and lymphoblastic cell lines were developed for each sample. The first comprehensive descriptions of human genomic variation (discussed in detail in Chapter 6) were produced using this panel. Since the establishment of the HGDP–CEPH panel, a veritable alphabet soup of genome diversity projects has been established, and more recently united under a single moniker – the Genome Aggregation Database (gnomAD; see below).

As more and more genomic data were collected for multiple humans, the need for a mechanism to collect and archive the information was recognized. Because the Human Genome Project didn't just produce the genome of a single human (multiple genomes were always the target of the project, as described in Chapter 6), researchers discovered a large number of the polymorphisms called single nucleotide polymorphisms (or SNPs). In the United States, the National Institutes of Health initiated a database called the Database of Single Nucleotide Polymorphisms (dbSNP). Then, to standardize the collection of SNP data, the International HapMap Project was created by NIH's National Human Genome Research Institute in 2002. The 1000 Genomes Project (1000GP) subsequently grew out of the desire to streamline the task of SNP data processing by standardizing the genomes that researchers used to discover SNPs in human genomes. Accomplished in two phases, and separated into several projects, the 1000GP (like the HGDP–CEPH) tagged the genomes of 2,500 humans as targets for extensive SNP searching. Also like the HGDP–CEPH, the 1000GP focused efforts on worldwide genomic variation by including people from across the globe, and from different ethnic groups. Figure 4.5 shows the location of the HGDP–CEPH and 1000GP samples.

Several more recent projects have been hatched since the establishment of the HDGP–CEPH and 1000GP. The Simons Foundation funded the sequencing of 260 human genomes from diverse geographic localities and of people from different ethnic backgrounds. The difference from the HGDP and other projects is the degree of coverage and accuracy of assembly of the genomes in the data set (260 genomes from 127 populations: 39 Africans, 23 Native Americans, 27 Central Asians or Siberians, 49 East Asians, 27 Oceanians, 38 South Asians, and 71 West Eurasians).

Direct to consumer (DTC) kits for genome analysis have also been a popular way to amass large numbers of human genomes with geographic and personal data. Literally millions of human genomes have been generated by these DTC efforts. Because eukaryotic genes are structurally complex, different parts of genomes have been separated into distinct components. Eukaryotic genes are comprised of regions that code for amino acids in their protein and noncoding regions interspersed among the coding regions, called introns. The regions that code for protein sequences are called exons and the collection of all exons in a genome is called the exome. In 2012, Daniel MacArthur, self-described

HGDP–CEPH

HGDP–CEPH panel numbering

AFRICA	WEST AFRICA	EAST ASIA	OCEANIA
1 Bantu	16 Bedouin	28 Han N	46 Melanesia
2 Mandenka	17 Druze	29 Han s	47 Papuan
3 Yoniba	18 Palestinian	30 Dal	
4 San		31 Daur	NATIVE AMERICAN
5 Mbuti	CENTRAL	32 Hezhen	48 Karitiana
6 Biaka	SOUTH ASIA	33 Lahu	49 Surui
7 Mozaabite	19 Balochi	34 Miao	50 Colombian
	20 Brahui	35 Oroquen	51 Maya
EUROPE	21 Makrani	36 She	52 Pima
8 Orcadian	22 Sindhi	37 Tuja	
9 Adygel	23 Pathan	38 TU	
10 Russian	24 Burusho	39 Xibo	
11 Basque	25 Hazara	40 Yi	
12 French	26 Uygar	41 Mongolia	
13 Northern Italy	27 Kalash	42 Naxi	
14 Sardinia		43 Cambodia	
15 Tuscan		44 Japanese	
		45 Yakut	

Figure 4.5 Map showing the location of the HGDP–CEPH and 1000GP samples. The table gives more precise geographic information.

"DNA parasite" started the Genome Aggregation Database Project. The "gnomAD" trawls databases for high-quality human genome data sets and aggregates the data into a single database (Figure 4.5). There are in general two kinds of genomic information stored by gnomAD – high-quality exome sequences and high-quality whole-genome sequences. To date, over 65 large human genome diversity projects have been "parasitized" for data and

1000GP

100GP panel numbering

AFRICA		
1 GWD	Gambian/Mandinka	113
2 MSL	Mende/Sierra Leone	85
3 EAN	Esan/Nigeria	99
4 YRI	Yorba/Nigeria	108
5 LWK	Luhya/Kenya	99

EUROPE		
6 FIN	Finnish in Finland	99
7 GBR	British/English Scot	91
8 CEU	Utah/ European	99
9 IBS	Iberian/Spain	107
10 TSI	Toscani/Italia	107

SOUTH ASIA		
11 PJL	Punjabi/Pakistan	96
12 GHI	Gujarati/Texas, US	103
13 ITU	Indian Telugu/UK	102
14 STU	Sri Lanka Tamil/UK	102
15 BEB	Bengali/Bangladesh	85

EAST ASIA		
16 CHS	Han Chinese/S China	103
17 CHB	Han Chinese/Beijing	105
18 COX	Chinese Dai/China	93
19 KHV	Kinh/Vietnam	96
20 JPT	Japanese/Japan	104

NATIVE AMERICAN		
21 MXL	Mexican/LA, USA	64
22 ASW	African/SW, USA	61
23 PUR	Puerto Rican	104
24 ACB	African/Caribbean	96
25 CLM	Columbian	94
26 PEL	Peruvian	85

Figure 4.5 (Cont.)

compiled into a single data set. GnomAD had two precursors – the Exome Aggregation Consortium (ExAC) and the Exome Sequencing Project (ESP), both dedicated to targeted sequencing of protein-coding regions of the human genome. Both targeted exome sequences and whole-genome sequences are included in gnomAD. To date, over 130,000 high-quality genomes have been archived in gnomAD, stratified into the following groups: African/African American, Amish, Latino/Admixed American, Ashkenazi Jewish, East Asian, European (Finnish), Middle Eastern, European (non-Finnish), and South Asian (Figure 4.6). Any genome not in the nine major groups is placed into "Other."

All genomic information mentioned so far is based on SNP profiles, referring to single base-pair changes observed between individuals. To explore the role of large-scale structural variants in the genome, the Human Genome Structural Variation Consortium (HGSV) was formed. In 2020, this consortium announced the production of 64 high-quality genomes from 25 different human ethnic groups from across the globe. This alphabet soup of projects, consortia, and archives, as well as the creation of the International Genome Sample Resource, has opened a seventh door to racial science. In Chapter 6 we discuss the methods that researchers use to analyze the genome-level data, and the ramifications of these analyses.

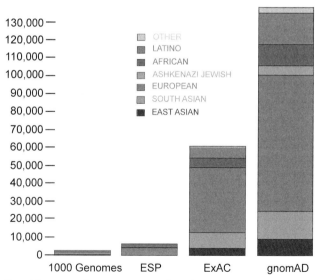

Figure 4.6   Geographic representation of samples in the Exome Aggregation Consortium (ExAC), the Exome Sequencing Project (ESP), and gnomAD.

Science, Racial Science, and Racial Nonscience (Racism)

We discuss the outcome of all this genomic data collection, and its analysis, in Chapter 6. But at this point it is useful to make a few distinctions, mainly between racial science and racial nonscience. Science is a broad and wonderful discipline that strives to explain the natural world. The great philosopher of science, Karl Popper, described science in its purest form as hypothesis rejection. In Popper's view, science only advances with the proposal of hypotheses about nature that are posed such that they can be tested by observation – and rejected if they don't fit the facts. Respecting this limitation is what makes scientific knowledge different from all other forms of knowing, and what allows us to increase our knowledge base and to make explanatory inferences about the natural world with confidence that any false beliefs or inferences collected along the way will at least in principle be susceptible to elimination. This notion of science has been occasionally challenged since Popper first proposed it back in the 1950s – and it is true that to make sense of things in science we sometimes provisionally have to make suppositions we can't immediately test – but by now the notion of hypothesis testing is deeply embedded in science. Many methods of inference are important tools in science, but we must be careful only to use those that allow us to test the hypotheses they generate. We must propose hypotheses that are as probing or as severe as we can. The more likely a test is to be rejected, the better the science. Testing simple hypotheses does not advance us much in science. On the other hand, if we use poor knowledge or misleading information to formulate a hypothesis (such as groupings of people) that has not been biologically defined, or is not susceptible to hypothesis generation or testing, then we find ourselves in the realm of "racial nonscience."

We have described the many doors that have been opened to give us our modern tools for studying genomic variation in humans. None of these open doors necessarily leads to racist conclusions, but passing through them may be conducive to research that is "racial" in its biological approach. To us, racialism in science includes any research program that uses racial assumptions, tests of racial hypotheses, and the use of race to make inferences about human variation. The Cann study allows us to reject the hypothesis of race existence, and we will show in Chapter 8 that genomic data are no better. Hence, nobody has been able to show that any "racial" division of humankind maps meaningfully on to any distinctive aspect of human biological variation, somatic or genomic, and such studies are from the outset unscientific. It might, of course, be thought useful to recognize "race" in medical or social contexts, and many studies have done so. But to the best of our knowledge, none of them has done so by properly characterizing those races in a biological context and thus remain unscientific, a subject we will return to in Chapter 8. Such studies are not necessarily "racist" in intent, but it is nonetheless true that science has regularly been hijacked to racist ends over the past couple of centuries. Even outside the political realm, race has been used to make generalized statements about groups of people based on unscientific and poorly developed ideas about human variation that in the end are only harmful to our understanding of the human condition. This is not science, and we can only deplore the continued use of race in science when it is neither warranted nor useful in explaining the natural world, and in turn in explaining our biology as a species.

At the opening of every door in the progression to modern racial science there has fortunately been pushback. While eugenics was a popular movement well into the 1940s, there were scientists who abhorred the idea, and clearly showed its scientific inadequacy. The blood group, allozyme, and early systematic work on human populations was also embraced in a racial context by many prominent scientists, but the racial implications of this work were vigorously resisted by Richard Lewontin and others in the 1970s. We will see in Chapter 6 how mitochondrial Eve and Y chromosomal Adam both advanced our understanding of human movement across the globe, but also resulted in unwarrantedly racist and unscientific conclusions until the results from molecular Eve and Adam were interpreted as haplotype evolutionary patterns. The final big door that opened to racial science resulted from the genomic revolution that started with the first draft of the human genome in 2001 and extends into the present. The development of human variation analysis panels, while not racist in conception, has led to several racist interpretations of these databases. We also deal with those interpretations in Chapter 6.

5 Variation among Genomes, and How Humans Took over the World

A big part of the story of our species, and of how variation is apportioned within it, involves how our ancestors spread over the globe. After all, if we had simply stayed in our place of origin in Africa and not ventured out, there would be no question that we are a single entity. That is because if our species had been restricted to a single location (as many others are), two things would have ensured that differentiation into separate entities would not have occurred. First, proximity would have ensured that mating would have been both possible and frequent. The capacity to mate is what gives species their integrity, because lack of mating allows for two species to diverge. The capacity to mate between two biological entities is the dominant factor in determining how a population will evolve. Second, all individuals of a very localized population are under very similar selective pressures. The amount of UV radiation from the sun, the food eaten, the parasites and diseases faced, and other environmental factors will be similar for most individuals in the population – quite unlike the case in a widely dispersed species such as our own. Indeed, *Homo sapiens* has taken a particularly adventurous path to its current worldwide geographic distribution; and tracing its history of spread has been an important part of understanding the phenotypic variation and the localized differences we see among its members.

Early Molecular Studies of Human Variation

Until about two decades ago, genetic researchers were limited to studying the migrations of humans using contemporaneous information. In 1983, Donald Tills and collaborators published a summary of what was known of human genetic variation in blood groups, and after electrophoretic variation and sparse DNA sequence variation had subsequently been added to the genetic variation arsenal, Arun K. Roychoudhury and Masatoshi Nei published in 1988 a comprehensive survey of human genetic variation, at 362 loci for 180 human populations. They summarized their data using population genetic approaches and distribution maps. In 1994, Luigi Luca Cavalli-Sforza, Paolo Menozzi, and Alberto Piazza produced a tome on human genetic variation that they titled *The History and Geography of Human Genes*. The book is a remarkable summary of the genetic and geographic patterns of hundreds of human genes, and hundreds of human populations. The gene targets varied, from blood groups to human leukocyte antigens (HLA loci), to allozymes, and to mtDNA variation. The criteria for defining the subject populations were very similar to those used by Tillis and colleagues and Roychoudhury and Nei, and are captured in the following quote:

> We have confined our analysis to aboriginal populations that were in their present location at the end of the fifteenth century when the great European migrations began. We have thus excluded Black Americans and all the recent colonizations of Caucasoid, Chinese, and Indian origins. We have also excluded all manifestly mixed populations; those stated to have 25% or more external admixture.

There are problems with these criteria (the assumption of no migration prior to the fifteenth century for one) that exist. The exclusion of mixed populations is also problematic and will skew observations and conclusions in the direction of more distinct differentiation between human populations. Nonetheless, Cavalli-Sforza and colleagues' methods of analysis were standardized for each locus and included (1) computation of genetic distances between populations; (2) construction of "trees" from those distances; (3) principal component analysis transposed onto maps; and (4) straightforward distribution maps. Many of the genes analyzed were accompanied by a map (Figure 5.1) depicting the geographic distribution of the alleles of the gene concerned. In all, 128 of these "maps" were included, representing 128 loci.

Other population demarcations were used that conformed to well-known ethnic groups on the planet, so the book can be characterized as "racial." The main tools for analyzing and visualizing the data involved tree-building

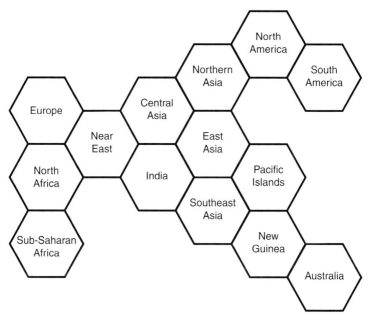

Figure 5.1   Cavalli-Sforza et al.'s 1994 map of the world showing the 14 regions of the globe that were used as general stratification boundaries. For appropriate loci the boxes would be filled with information about allele frequency for that locus.

and clustering, two approaches that are now widely used in genomic human variation studies.

The Y-Adam and mt-Eve Stories

The accumulation of data categorizing the thousands of haplotypes that we know of today began in the 1980s. These early studies were based on polymorphisms discernable by restriction fragment length polymorphism (RFLP) analysis (see Chapter 4), and yielded a sound foundation for future work and for the discovery of thousands of variants (haplotypes) that can be grouped into haplogroups. Currently, it is simple and inexpensive to sequence the whole mtDNA genome, and to obtain thousands of SNPs from Y chromosomes. Blood or buccal samples are obtained from individuals, and their DNA is isolated. The DNA is then sequenced either using a chip, or directly with "short reads," and then assembled. Polymorphisms (and hence new haplotypes) are discovered using bioinformatic techniques that compare the new sequence to those already in the database, and the results can be displayed in several ways. One popular approach is to create pie diagrams that summarize the frequency of different haplotypes for a geographic area. Since the topology of the tree that represents these haplotypes is well known, tree-building methods have by now become somewhat redundant for mtDNA haplotype studies focused on racial differentiation. In other words, the tree doesn't tell us much more than what was already known with respect to patterns of human mtDNA differentiation.

Researchers have also used other ways of presenting haplotype data. The haplotype and haplogroup data are organized around a very simple but sound nomenclature. Each haplogroup is given a letter name (A, B, C, etc.). The letter names of the haplogroups go from A to Z. Numbers are added to the letter names when slight variations occur within a haplogroup. Haplogroups have been named in the order of their discovery, meaning that haplogroup A was discovered and characterized before haplotype F. This naming scheme means that the alphabetical ordering says nothing about the actual genetic relationships of the haplogroups.

Figure 5.2 is an example of an approach that assumes certain haplotypes are descriptive of certain geographic areas (for instance, haplotypes A, B, and C are found mostly in the Americas). The different population groups in the 1000 Genomes Project (1000GP) are then categorized based on their diagnostic haplogroup polymorphisms, using a form of clustering that employs predetermined geographic boundaries to organize data. This essentially racial "stratification" allows for easier data visualization of haplotypes into five geographic areas – Africa, South Asia, East Asia, America, and Europe. One very precise analytic tool involves characterizing the haplotypes according to whether they are "diagnostic" or not. A diagnostic haplotype is a marker, in this case a distinctive haplotype that is fixed (it is the only haplotype) in one geographic area.

CONT	Pop.	AFRICA						EAST ASIA						EUROPE									ASIA		AMERICA		
		L0	L1	L2	L3	L4	L5	D	F	G	N	Y	Z	H	I	J	K	T	V	W	X	U	M	R	A	B	C
AFRICA	ESN	7	20	27	43	2																					
	GWD		15	48	42	2																					
	LWK	18	8	12	47	5	10																				
	MSL	2	17	38	24	1																					
	YRI	5	17	38	47	1																					
EAST ASIA	CDX							12	25	5	8	1	1									24	14	7	9	5
	CHB							23	16		10		3									19	6	6	12	5
	CHS							23	16	2	9		4									19	11	6	16	2
	JPT							39	6	11	4											14			14	1
	KHV							2	27			1										33	11	1	21	
EUROPE	CEU													51	1	8	3	10	3			18					
	FIN													37	2	7	6	3	5	2		36					
	GBR													39	3	10	5	10	2	2	2	18					
	IBS							1			2			53	1	4	7	8	2	2	5	18		1			
	TSI													54		8	9	13	4		1	15					
ASIA	BEB							2	1					7		2				2		11	58	8	1		
	GIH													7	1	1		2		2	2	15	41	32	1		
	ITU													2		1		4		5		14	61	13			
	PJL									2				7		1		3		2	1	11	55	11			
	STU								3		3			12		1						14	50	22			
AMERICA	ACB	4	21	38	26			1						1		1							1		1	1	1
	ASW	7	16	14	24	1		2															1		1		
	CLM	1	2	2	4			9			1			1			2	2					1		40	33	6
	MXL							13															1		25	15	9
	PEL		5	4	2									5					1						14	40	7
	PUR	2	2	3	13									2		6	1	1		2		4	1		38	7	25
	WORLD	46	123	224	272	12	10	127	94	20	37	2	8	278	8	50	34	56	16	25	11	136	377	129	141	178	70

Figure 5.2 Example of using diagnostic haplotypes to characterize geographic grouping using the 1000GP data set. Geographic areas are listed from left to right in the first row and first column of the figure, while the mtDNA haplotypes are listed in the second row. The different populations used in the analysis are listed in the second column of the figure, and can be found in Figure 4.3. The numbers in the boxes of the figure indicate the number of individuals for each of the categories. So, for instance, the 7 in the third column and third row represents the number of individuals who have the L0 haplotype and are from the ESN (Esan, Nigeria) population. This approach assumes that certain haplotypes are descriptive of certain geographic areas (for instance, haplotypes A, B, and C are found mostly in the Americas and only sparingly in Asia).

A hypothetical example of diagnostic analysis for five regions is shown in Figure 5.3. First, look at haplotype A3. Note that it is present only in East Asian individuals. While this haplotype is not fully diagnostic, it is what is called "private" to East Asia. Next note that all three haplotypes in haplogroup B are found only in European individuals. This means that haplotypes B1, B2, and B3 are together diagnostic for Europe. It also means that haplogroup B is diagnostic for Europe, while haplogroup A is predominant in East Asia, haplotype C is predominant for Africa, haplotype D predominates for America, and

			HAPLOGROUP		
HAPLOTYPE	A 123	B 123	C 123	D 123	E 123
EAST ASIA					
1	551		1 1		
2	551				
3	551		1 1		
4	551			5	
EUROPE					
1		555			
2		555			
3		555			
4		555			
AMERICA					
1	11			55	
2	1		1	555	
3	1		1	555	5 5
4	1			5	
AFRICA					
1	1		555	1	111
2	1		5 5	1 1	111
3	1		555		1
4	11		555	111	11
SOUTH ASIA					
1	11		111	1	555
2	1		111	1 1	5 5
3	1		111	1	555
4	1		111	111	5 5

Figure 5.3   A hypothetical example of diagnostic analysis for five regions and three haplotypes per area. The numbers in the boxes indicate the number of individuals with that haplotype.

haplogroup E predominates for South Asia. These latter area-specific dominant haplogroups can be used by geneticists as ancestrally informative markers (AIMs), although with some caveats. We will address AIMs from nuclear sequences later in this chapter.

Likewise, in the table shown in Figure 5.2, haplotypes are given the general classes African, East Asian, Asian, European, and American, based on the general distribution of haplotypes. There are three haplotypes that are entirely private to an area: L5 is private to Africa, while Y and Z are private to East Asia.

Other population genetic measures may also be used to characterize the compositions of haplogroups for both mtDNA and Y chromosomes. Researchers can ask questions about the genetic isolation of populations, or about how the variation in the populations is apportioned, using a wide range of approaches that are discussed later in this chapter. The diagnostic and population genetic approaches mentioned here are important, as they are also the major ways in which botanists and zoologists determine species boundaries in animals and plants. It should also be pointed out that the haplogroups are not intrinsically "racial": What makes a study racial is the application of the geographic boundaries that researchers place on their data sets. Figures 5.1 and 5.2 are good examples of this approach, and both seem to us to reject quite soundly any hypothesis of definable races in their data sets, because different haplogroups are found in several of the hypothesized groups. Another way to say this is that there are no fixed diagnostics for any of the hypothesized groups.

It should appear fairly obvious that adding more haplotypes, or adding more individuals, might result in the obscuring of the private and diagnostic haplotypes. For instance, if a new study were done and added to the hypothetical one in Figure 5.3, and any one of the new individuals from East or South Asia, Africa, or America were found to have a haplogroup B haplotype then the pure diagnostic nature of the B haplotype would have to be rejected. Because of this, it is a good idea to ask how complete our knowledge is of human mtDNA haplotypes. Figure 5.4 shows the accumulation of novel haplotypes with time (the accumulation of haplogroups would be similar) for both Y chromosomes and mtDNA genomes. Both curves show gradual accumulation of unique haplotypes with time, and a significant rise in the number of haplotypes

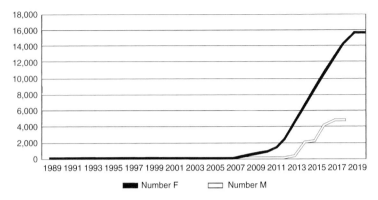

Figure 5.4   Graph showing the accumulation of haplotypes with time. The black line indicates female mtDNA haplotypes and the white line indicates male Y chromosomal haplotypes.

found between 2011 and the present (due to the introduction of next-generation DNA sequencing technology). Both markers also appear to be leveling off with respect to the accumulation of new haplotypes, and hence of new haplogroups, so that it is unlikely that the inferences we explain above will change in future.

Paleogenomics

Genetic materials from long-dead hominin individuals have provided a unique window into our recent evolutionary history under the rubric of "paleogenomics." As we pointed out in Chapter 4, researchers have focused both on proteins and on the DNA that encodes them. Proteins and DNA are both linear molecules, but whereas proteins are long by molecular standards, with the average amino acid number per protein being on the order of 1000 residues, DNA molecules are massively long. They make up our 23 individual chromosomes, the smallest of which is about 20 million residues in length, and the largest 10 times that. Both DNA and protein molecules can be sequenced using modern techniques, but most researchers focus on DNA not only because it is technically easier to isolate and sequence than proteins, but also because if you have the DNA sequence you need only to translate it, via the genetic code, to obtain the protein sequence, whereas the opposite is not possible due to the nature of the genetic code.

While DNA can be extremely fragile in humid and hot conditions, it can survive in fossils and subfossils for long periods of time under the right conditions (usually very cold and arid). There are three kinds of DNA markers in our genomes that paleogenomics researchers routinely use: a maternal component (mtDNA), a paternal component (Y chromosome DNA), and a recombining nuclear component (most of our DNA in the nuclei of our cells). These three kinds of markers can give different signals of relatedness and ancestry. Due to the difference between clonal inheritance and recombining genetic elements, researchers have developed divergent methods to follow the genetic changes in paternal and maternal lineages, versus the recombining portion of the genome. Branching diagrams (called phylogenetic trees) can be relatively straightforwardly generated to reconstruct the evolutionary history of these two kinds of clonal markers. In contrast, the 98 percent of the genome that recombines during reproduction produces a more intricate history that greatly complicates reconstructing evolutionary hierarchies.

Sequencing the genomes of living individuals is easy due to the availability of high-throughput next-generation sequencing (NGS), but sequencing from fossil specimens is technically difficult due to natural degradation and contamination of fossil samples over time. Fortunately, these problems have been overcome through very clever technical approaches mostly developed by Svante Pääbo and his colleagues at the Max Planck Institute in Leipzig, Germany. The major technological breakthrough was a combination of NGS techniques that allow billions of small, degraded fragments of DNA to be sequenced using computational methods that take the small fragments and assemble them into longer pieces. As a result, under favorable conditions genome sequences from long-dead hominins can now often be generated with confidence and relative technical ease. Pretty complete genomes from people dead for hundreds of years are commonplace, and one 430-kyr-old (1 kyr = 1,000 years) fossil from Spain has yielded mtDNA fragments. Currently, several thousand "ancient" human mtDNA genomes have been sequenced, along with nearly 1,000 or so ancient Y chromosomes. Nuclear genomes are more difficult to obtain, but over 1,000 ancient nuclear genomes have also been sampled and sequenced in the last decade. Research has focused on European lineages; about half of the sequenced mtDNA genomes and over two-thirds of the sequenced Y chromosomes are from individuals found there. The great majority of fossil mtDNA and Y chromosome genomes have been sequenced from fossils under around 10 kyr old, but the numbers of older ones are growing.

Analyzing hominin paleogenomes is the pioneering work of several groups of researchers, almost all of them connected at one time or another to the Max Planck Institute. These researchers have focused on three major endeavors related to human evolution: (1) coarse-grained ancient migration events out of Africa and into the five habitable continental regions of the planet; (2) more fine-grained migration patterns within each continental region; and (3) the timing of migration events at both a coarse level (i.e., between and within major areas of continents) and fine-tuned level (i.e., highly localized areas). It is possible to do this by using molecules both as evidence and as a clock. While DNA sequences do not change in a clocklike fashion like the radioactive isotopes geologists use to date their rocks, there are clever methods based on assumptions about fossil ages and rates of sequence change that can be used to determine the time ranges of migration events using the changes that occur among genomes.

Most of what we know about coarse-grained patterns of divergence (intercontinental movement) of humans comes from mtDNA and Y chromosomal DNA sequences in living humans. But the addition of ancient mtDNA and Y chromosomal genomes has expanded this knowledge substantially, and the haplogroups that we discussed in Chapter 4 can be illuminating here. About 5,400 or so mtDNA haplotypes and 4,600 or so Y chromosomal SNPs have been identified so far. All mtDNA haplotypes can be boiled down into 38 mtDNA haplogroups, and the Y chromosomal SNPs into 48 Y chromosomal haplogroups. Both sets of haplogroups generally coincide with geographic locations.

The haplogroups themselves are used to detail relationships through phylogenetic analysis, and the relationships in the trees that are generated by such analyses can tell us which haplogroups are concentrated in which geographic regions of the globe, much like the clustering method illustrated earlier in this chapter (Figure 5.3), or the pie diagram approach. The difference is that a tree can reveal the order in which movements happened. When we root the trees of both human mtDNA and Y chromosomal haplotypes with their chimpanzee equivalents, it is evident that African haplotypes branch off first (Figure 5.5). This indicates that our species originated in Africa, something long known from fossil evidence. Africa also harbors the greatest amount of haplogroup diversity anywhere on the planet, also suggesting deep roots. The older a lineage is, the more diversity it will show because the longer time the lineage exists allows for more opportunities for diversity to arise. The mitochondrial data suggest the existence of three major lineages in Africa (L1, L2, and L3 haplogroups), and the deepest-branching living populations in these three major haplogroups include the central African rainforest hunter-gatherers, southern African Khoi-San, and the Hadza of Tanzania.

The major inferences that can be made by adding fossils to the existing haplogroup structures established from living humans involve placing geography in a temporal framework. Cosimo Posth and colleagues sequenced the whole mtDNA genomes of 55 ancient humans from Europe who ranged in age from 7 kyr to 35 kyr. From their ages some major inferences about the colonization of Europe were possible. A major L3 haplogroup dispersal out of Africa some 60–80 kyr ago was followed by splitting off of the non-African M and N haplogroups, and the two haplogroups derived from M and N – respectively, R and U – at 60 to 50 kyr ago. These latter haplogroups are found among living humans in the Gulf oasis region of the Arabian Peninsula. The analysis suggests that haplogroups M, N, R, and U then dispersed eastward into the South Asia region, and westward to Europe (with some back-migration to North Africa). The situation in Europe over the last 40 kyr is complex due to a massive amount of population movement in this area, although Posth and colleagues were able to infer that there was a severe reduction of human populations in Europe at the Last Glacial Maximum (LGM) between 15 and 19.5 kyr ago. The majority of humans surviving this event had U2-like haplotypes. Surprisingly, by a few thousand years later almost all haplotypes sampled were U5, demonstrating a drastic population turnover in this period of time. To fill in the time range between 7 kyr ago and the present, Guido Brandt and colleagues determined the mtDNA haplotypes of 364 prehistoric individuals from Central Europe. Their results indicated several other marked shifts in the demography of human populations in Europe over the last 7 kyr. Figures 5.6 and 5.7 show

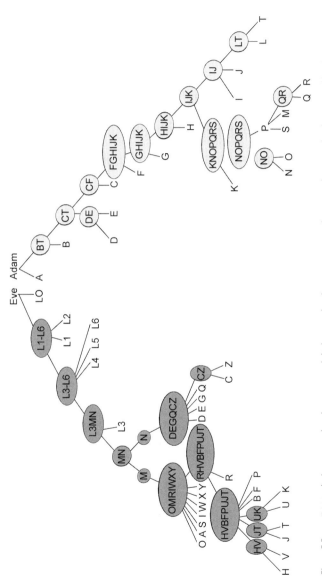

Figure 5.5   mtDNA (dark gray) and Y chromosomal (light gray) haplotype trees. For the general geographic location of haplotypes, see Figure 5.2.

Figure 5.6   Diagram showing the kinds of sequences encountered during the ascertainment process.

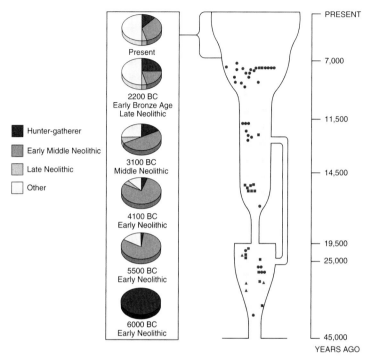

Figure 5.7   Shifting demography of ancient human populations in Europe, based on clonally inherited markers. The pie diagrams represent the proportion of haplotypes. Triangles represent M haplotypes, squares represent U haplotypes (U2, U3, U4, U7, U8, U9), and dots represent R and U5 haplotypes. Far left shows the percentage of the indicated haplotypes in Europe-wide surveys of individuals from the indicated times. There are several turnovers of haplotypes in the figure. From 6000 BC to 5500 BC there was a dramatic turnover from the hunter-gatherer haplotypes to what are called early Neolithic haplotypes, and there was another large turnover between 4100 BC and 2200 BC from early Neolithic to the haplotypes that currently characterize the European continent. The diagram shows a demographic model from 45 kyr ago to the present. Note the constriction in population at the LGM, and a turnover event at about 14.5 kyr ago.

the fluidity with which human populations have changed over time in this area of the world.

The autosomal portions of ancient human genomes can also be important in clarifying past population movements. But before we delve into that subject, we need to describe how the genomes of ancient and extant humans are obtained. Technically, characterizing mtDNA and (in some ways) Y chromosomal DNA is relatively simple compared to analyzing the autosomal component of the human genome. Because crossing over of genetic material in autosomes is common, autosomal segments get mixed up, making it harder to extract the genealogy of autosomes. Clonal markers do not recombine readily like autosomal markers, and so the genealogy is easier to interpret. Consequently, researchers have developed some very clever approaches to obtaining the genome sequences of humans, both alive and long-dead.

Genomic Shortcuts

Autosomal genomic studies have been essential to understanding fine-grained population dynamics, and these studies have been made possible by several new high-throughput methods. While it has become increasingly easy to fully sequence human genomes, many researchers nonetheless use shortcuts that expedite the process, many depending on the fact that most of the bases of the three billion we all have in our genomes are identical from one individual to the next, and that it is thus a waste of time and effort sequencing them. Enter "targeted sequencing," an approach that focuses exclusively on those positions in the genome that are variable, and thus informative for human variation studies. For any population of interest this can be done by lining the sequences up so that the positions in the sequence can be compared easily (in a process known as alignment). The variable sites (SNPs; see Chapter 4) are used to separate the relevant stretches of DNA from the rest of the genome being sequenced. Once the targeted sequences are separated, they can be sequenced using NGS approaches. Alternative "array" approaches use a different technology that hybridizes target DNA to probes bound to a microchip. Targeted sequencing approaches allow for replication sequencing at about 100× coverage (where "coverage" refers to the number of unique reads for any given SNP in a single sequencing experiment). In a typical study,

Panel name	Focus	Number of SNPs
Geno2.0 Genographic Project	Ancestry	150,000–700,000
deCODEme	Disease/ancestry	Over 1,000,000
23andMe	Disease/ancestry	Over 600,000
IDT panel	Disease	Over 1,000,000
Illumina HuHap 650 K	Ancestry	Over 650,000
Skoglund Ancestry 850 K	Ancestry	Over 850,000
CleanPlex	Disease	Over 1,000,000
CRI Genic Panel	Ancestry	Over 500,000

Table 5.1 Some examples of human ancestry- and disease-targeted sequencing panels

several hundred thousand SNPs will be assayed by one of these methods. The SNP panels used to do the separating are commercially available; some are proprietary, as listed in Table 5.1.

How the panels are engineered is of interest in the context of race. With millions of SNPs to choose from, narrowing the number of SNPs down to a manageable level is an important technical aspect of human variation studies. If one is studying disease, then perhaps the best way to design a diversity panel is to focus on those genes that cause disease and to look only at highly reduced subsections of all SNPs. If one's focus is ancestry, it is desirable to capture as much "meaningful" variation as possible for the least effort possible. In either case, a shortcut is necessary, and the key thing here is how the shortcut is designed. The process of designing a human variation or human origins panel is called "ascertainment," and, like any shortcut method, it is a tricky business.

The backbone of the ascertainment procedure is a fully sequenced and well-characterized human reference genome. Other individuals carefully chosen to represent as many lineages and geographic locations as possible are then sequenced, at what is called "low coverage," to compare to the reference sequence that has been determined at very high coverage. The trick is to find the lowest number of SNPs that allow the researcher to obtain overlapping data at a given coverage level. This overlap is required because at least two reads of the data are required to assess whether a SNP is variable. The choice of individuals for the low-coverage sequencing depends on the specific study, and here is where the word "meaningful" comes in. Meaningful for a coarse-grained analysis of intercontinental populations is different from meaningful for a study of fine-grained populations, say in France. Fortunately for the researchers who do this kind of work, large databases of sequences already exist (Chapter 4) in the guise of the HGDP–CEPH project, the 1000GP, and other projects like the Simons Genome Project.

The sequences from the reference individual and individuals from the database (also called the "ascertainment group") are then matched, and a set of predetermined rules is used to *ascertain* the SNPs that the researcher thinks will be useful in a study. Figure 5.8 shows the process of ascertainment, where the rule established is that at least 2× coverage (at least two copies of the potential SNP must exist in the data) is needed to ascertain a SNP. More stringent ascertainment criteria can be used. This process is only part of what "meaningful" might mean. Imagine that you are studying the variation patterns of all people on the planet. If you are looking at the ancestry of human lineages, your choice from the already existing databases will be based on the geographic diversity relevant to the question you are addressing. For this global assessment you would choose individuals in the database who represent a broad array of human genomes from all over the globe. If, on the other hand, you are interested in how populations evolved in France, you would increase the probability that you will ascertain large numbers of SNPs for your study by selecting those genomes from the database that are most closely related to people in France. Researchers using this approach must assume that the individuals of choice have well-defined ancestries. Of course, that assumption may not always be valid.

The first sequence variant in Figure 5.6 is not ascertained, because only one copy of the fragment is found in the data, violating the 2× criterion. The last potential SNP fragment on the far right is also not ascertained, because it is not variable in the two new sequence reads. The two potential SNPs in the middle (downward arrows) *will* be ascertained, because they are variable in the new reads. Again, whether these regions are variable in the first place is entirely dependent on which sequences from the database are used for the ascertainment.

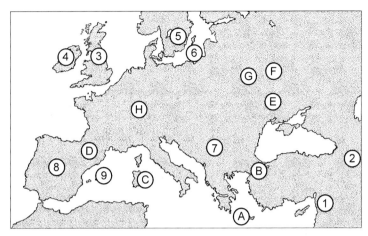

Figure 5.8   Demographic patterns of ancient human populations in Europe based on clonally inherited markers. The numbers on the map refer to specific demographic events described below: (1) Paleogenomes indicate that present-day Lebanese are closely related to ancient Canaanites; since the Bronze Age, there has been substantial genetic continuity in the Levant. (2) Ancient genomes of Fertile Crescent fossils indicate multiple, genetically differentiated hunter-gatherer populations who adopted farming in southwestern Asia. (3) Ancient Roman fossil genomes were generated, and showed affinity with modern British Celtic populations that are significantly different from modern-day eastern English people. (4) Several fossil genomes generated from specimens from Ireland indicate a substantial influx of early farmer genomes to the island. (5) Fossil Scandinavian genomes suggest a postglacial migration from the south, and later one from the northeast. (6) Several Baltic-area human fossil genomes suggest that the first Scandinavian farmers arose by migration from Anatolia about 1,000 years ago. (7) Multiple fossil genomes from the Balkans demonstrate that southeastern European genomes are genomically distinct from Western European genomes. (8) Several paleogenomes from the Iberian Peninsula suggest that early Iberian farmers were significantly genomically different from Central Europeans, suggesting two independent points of Neolithic migration. (9) Ancient Mediterranean paleogenomes cluster with all previously analyzed modern-day Europeans. (A) Ancient southern Greek paleogenomes indicate that Minoans and Mycenaeans were genetically similar. (B) Aegean paleogenomes show striking similarity with genomes from across Europe, indicating a direct genomic connection of the Mediterranean and Central Europe. (C) Paleogenomes from Sardinia confirm a Mesolithic occupation of the island. (D) Paleogenomes from Germany, Hungary, Spain, and Russia indicate closer affinity of the former three, as

Ascertainment sets a baseline for variation studies in human genetics and human origin studies. If it is done incorrectly, or in a biased fashion, problems can arise. What happens if a researcher isn't entirely careful and unbiased? Joseph Lachance and Sarah Tishkoff suggested that biased sets of pre-ascertained SNPs can cause genotyping arrays to produce erroneous inferences. One kind of bias is called "minor allele frequency bias." This can result in overrepresentation of SNPs that have high minor allele frequencies (frequencies where the minor allele is not rare), and in underrepresentation of SNPs that have low minor allele frequencies (frequencies where the minor allele is very rare). The second kind of bias concerns the number of individuals in an ascertainment group or subpopulation. This parameter will influence the lower limit of frequencies of alleles in populations, so that SNPs that exist in low frequencies are unlikely to be observed in an ascertainment group because it will be unlikely to find individuals in the sample that have those.

Since clustering is one of the major approaches used when analyzing human variation data, a look at how it is impacted by ascertainment bias is warranted. Low-frequency alleles may be the result of recent mutations that are limited to specific geographic areas simply because they have not had time to be spread around. If the ascertainment bias is against these SNPs, then extra information on geographic clustering of alleles for the SNP will be excluded from the study and the clusters obtained will be fewer and inaccurate. If, on the other hand, those low-frequency SNPs were biased for a panel, then more geographic clustering will be inferred than warranted.

Because of these uncertainties, the answer to any question one might ask without a fully objective set of SNPs may be biased, because while the SNPs selected for the array or the targeted data set may be chosen for reasons that

Caption for Figure 5.8   (cont.)

compared to Russia. (E) Paleogenomes from the Eurasian steppes suggest that this zone can best be described as a mixture of Yamnaya genomes and East Asian genomes. (F) Several paleogenomes from Russia indicate that Caucasus hunter-gatherers were members of a distinct ancient clade that split from western hunter-gatherers around 45 kyr ago. (G) Sequencing of many paleogenomes from the Steppes of Eurasia suggest that they emerged from the same ancestral gene pool as early farmers in other parts of Europe.

are fully in line with the initial research question, they may not be applicable to subsequent research questions. What this means is that we need to be very careful about the actual research questions we ask, and about how we approach answering them. This is key when we are looking at questions about human origins and history, for it is imperative to exclude any "ascertainment bias" resulting from inadequacies in the ascertainment process. To drive this point home, consider the establishment of what human geneticists call "ancestral informative markers." In human genomics these markers are supposedly fixed SNPs that differentiate between specific populations. To us, however, AIMs are the result of an extreme ascertainment process.

The View from the Autosomes

Existing analyses of mtDNA and Y chromosomal data have pretty much nailed down the history of our species at a coarse level. More recent studies are usually focused on specific continents, or even on more specific geographic regions within a continent. Here we summarize several of these studies.

Africa: Paleogenomicists are actively addressing issues relating to when and how many times early modern humans exited Africa, and if there were any back-migrations. Using a molecular clock approach to estimate the time of major divergence and movement of *H. sapiens*, Svante Pääbo and colleagues arrived some years ago at an estimate of up to 180 kyr, in pretty good agreement with the paleontological evidence. The most recent estimate of this event in 2016 has pushed the date back to 260–350 kyr by generating sequences from several fossils from KwaZulu-Natal, South Africa and comparing these to the major lineages of living people with African ancestry. This new date would almost double the divergence time of our species from a currently unknown predecessor species, and it is attended by uncertainty. The subsequent watershed event was the move away from hunter-gatherer societies to more pastoral-agricultural ones, and paleogenomic studies indicate that Bantu-speaking herders and agriculturalists moved out of their montane homelands in current-day Nigeria and Cameroon some 4 kyr ago, to eventually occupy most of sub-Saharan Africa (Figure 5.9). The paleogenomic data suggest that, during this pastoral and agricultural spread, hunter-gatherer

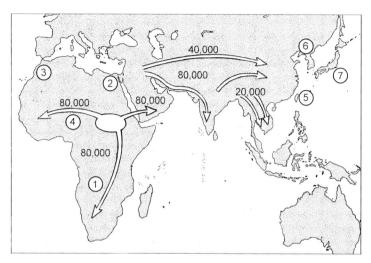

Figure 5.9   Map showing the migration of humans out of Africa and into Asia.

human populations in this vast stretch of Africa interacted in various ways to produce current-day populations through wide admixture.

Asia: It was generally believed until recently that *H. sapiens* initially entered Asia about 60 kyr ago. A second view is that *H. sapiens* movement into Asia occurred in two waves, an older one (90–120 kyr ago) and a younger one (40–60 kyr ago). The most recent paleogenomic data support the latter hypothesis, also known as the "Two Layer Model." It is the younger wave that appears to have contributed most to the current genomic structure of Asian populations. Several ongoing studies are unraveling the peopling of other Asian areas at a finer level (Figure 5.10). For instance, researchers have shown that genomes from 500-year-old to 4,500-year-old hunter-gatherer fossils from Laos and Malaysia are closely related to present-day Onge hunter-gatherers from the Andaman Islands, and that contemporary East Asian genomes are more closely related to Neolithic fossil genomes found in Southeast Asia. Clearly,

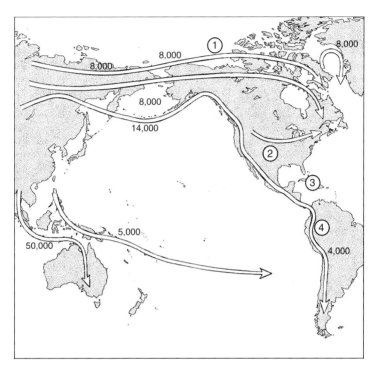

Figure 5.10   Map showing the migration of humans out of Asia and into the Western Hemisphere.

the Two Layer Model is supported by these data. Other continental *H. sapiens* populations were greatly impacted by these initial waves into Asia.

Australia and the Pacific: While no fossil genomes (clonal or recombining) are available from Australia (one 100-year-old aboriginal genome has been generated from a tuft of hair), the evolutionary history of *H. sapiens* on that continent has been examined using about 75 whole-genome sequences from Pama-Nyungan aboriginal Australians, and 25 modern-day Papuans. *Homo sapiens* in both regions diverged from Eurasians between 60 kyr and

100 kyr ago, with the Papuans splitting off 25–40 kyr ago. Australian aboriginal people began to spread further between 10 kyr and 32 kyr ago. The Pacific islands were peopled much more recently, and from very different source populations. About 25 percent of the genomes of people in the South Pacific today come from Papuan sources, and the remainder from East Asians. The question becomes one of when the Papuan contribution was made: before or after the movement of humans into the islands of the South Pacific? By examining ancient genomes from 3,000-year-old fossil humans from Vanuatu and Tonga, researchers were able to show that the mixing of Papuans seen in contemporary Pacific Islanders began after their migration to Oceania, because the fossil genomes show no trace of the Papuan genomic content.

The Americas: Questions of two kinds have been at the forefront of research on the peopling of the Americas. The first kind concerns when and how the first people arrived, and the second kind concerns the fine-grained movements of people within the New World. It was fairly clear from paleontological, geological, and archaeological data that the Americas had been entered by people from Siberia at least 14.6 kyr ago, and perhaps as early as 23 kyr ago. Details of what happened after the migration event are just now being teased apart by adding paleogenomic information. Some researchers had proposed that this Siberian migration event was complemented by assimilation with migrants from Asia or the South Pacific, but this view of the peopling of the Americas can be rejected by examining the genome sequences of a 12-kyr-old human from a Clovis burial site in Montana, and of the 9.5-kyr-old Kennewick Man remains from Washington State. Both paleogenomes indicate direct Native American ancestry and help to reject the Asian/South Pacific migration hypothesis. Genomic data have also been used to show a basic bifurcation among Native Americans, with genomes that diverged from each other about 13 kyr ago giving rise to the Athabascan-speaking and Amerindian-speaking peoples. Paleogenomics has also revealed the existence of a second major migration event around 4 kyr ago. This involved now-extinct Paleo-Eskimo cultures that were followed by Inuit and Native American expansions into the New World Arctic. These data reflect the rise and fall of a population of people in the Arctic that was largely isolated from other populations for about 4 kyr.

Europe: Much of the peopling of Europe has already been discussed, so we give only a short summary here. Paleogenomics has shown clearly that three ancient genomic components shaped modern-day Europeans. The earliest settlers of Europe were hunter-gatherers who started to arrive around 43 kyr ago and continued to spread throughout the early Paleolithic. As noted earlier, a significant genomic turnover around the LGM at around 19 kyr ago suggests that they probably made little contribution to the genomes of existing Europeans. Neolithic farmers originating far to the east in the Fertile Crescent started trickling into Europe around 9 kyr ago, colonizing the Iberian Peninsula 7 kyr ago and Scandinavia and Britain about 6 kyr ago. A final wave came during the late Neolithic and early Bronze Ages, when people of the Yamnaya culture from the Eurasian Steppes migrated into Europe about 4.5 kyr ago. While all of this might seem quite complicated, Rasmus Nielsen and colleagues summarize by noting that modern-day European genomes show genomic "contributions of hunter-gatherers to the recolonization of Europe after the [LGM], the migration of Neolithic farmers from Anatolia to Europe and the late-Neolithic period and Bronze Age migration to Europe from the east."

Neanderthals, Denisovans, and a Third Mystery Lineage

It has been possible to extract and characterize both mtDNA and nuclear DNA from a limited range of nonmodern hominin fossils dating from the relatively recent past. Those extinct hominins include the well-known and large-brained *Homo neanderthalensis*, which flourished widely in Europe and western Asia after about 200 kyr ago, before becoming extinct after modern humans moved into their territory about 43 kyr ago. They also include the Denisovans, a shadowy group first recognized in 2010 purely from genomic information furnished by an undiagnostic finger bone recovered from the Denisova Cave in Russia's Altai Mountains. That original specimen is now believed to be around 60 kyr old and, together with a handful of other fragments from Denisova, it has now been joined by a partial lower jawbone from a high-altitude site on the Himalayan Plateau that is dated to about 160 kyr.

To date, there are about 20 Neanderthal specimens and three Denisovan individuals for which researchers have paleogenomic data, plus there is a single very divergent paleogenome from the Spanish site of Sima de los Huesos that has yielded fossils of a Neanderthal precursor that is as yet unnamed. This unprecedented number of paleogenomes has permitted the extrapolation of some very interesting evolutionary dynamics. For example, Petr and colleagues were able to compare the Y chromosomal evolution of Neanderthals and Denisovans. This was no easy task, as the majority of specimens sequenced to date from these two human species have been female. The researchers added up Y chromosome sequences to end up with two Denisovan and three Neanderthal Y chromosome complements. With the caveat that this is a small sample size, their analyses suggest that the Denisovan lineage (which should actually be acknowledged as a species) initially split from a lineage that harbored the ancestor of *H. sapiens* and *H. neanderthalensis*, and that *H. sapiens* and *H. neanderthalensis* diverged from each other significantly later. The Y chromosomal and mtDNA patterns are very similar, and if accurate this would mean that both mtDNA and Y chromosomes of Neanderthals were replaced by *sapiens* chromosomes.

On the autosomal side of things, paleogenomic studies using Neanderthals and Denisovan fossils yielded evidence of interbreeding in a seminal paper published in 2010. The work associated with the initial draft of a Neanderthal genome suggested that 1–2 percent of living *sapiens* genomes with Asian and European ancestry came from Neanderthal introgression. The initial timing of this interbreeding event with Neanderthals was 47–65 kyr ago. Recent Neanderthal and Denisova paleogenomes from the Altai Mountains in Siberia indicate a much earlier interbreeding event at about 100 kyr ago. And to add to this interbreeding event, paleogenomicists claim that at least three or four other instances of interspecies hanky-panky must have occurred. *H. sapiens* and Denisovans have apparently interbred in two distinct episodes, with people in Southeast Asia showing derivation of a significant part of their genomes from such interbreeding events. Denisovans and Neanderthals overlapped in their distributions, often even sharing microhabitats – even Denisova Cave itself – and so it is no big surprise that paleogenomics has uncovered evidence of interbreeding between these two lineages. And finally, when researchers examined the draft genomes of Denisovan specimens, they realized that a significant part of this genome contained chunks of a genome of a yet to be found "ghost" lineage that had intermingled with Denisovans.

It is stunning that paleogenomics has only been around in earnest for a decade, yet has already changed the landscape of our understanding of the origins, divergence, migration, and structure of our own species. Two things are certain from this revolution. First, more and more of paleoanthropology will turn to genomic analysis of fossil remains. It is a field within paleoanthropology that is not simply a flash in the pan, and that offers the prospect of revealing and unraveling many previously difficult-to-discover events in our past. The second certainty is a biological one. As we examine more and more of our genomic heritage and ancestry, using the genomes of both living and ancient people, we will continue to see that migration and interbreeding are and have been major themes of our existence on this planet. Understanding this tendency to migrate and interbreed is essential not only to our understanding of our own divergence and variation, but directly impacts how we view our species as a biological and behavioral entity.

6 Clustering and Treeing

A description of some of the methods used in racial genetics and genomics is essential if we are to understand why claims have been made that modern genomics establishes the existence of race. In this context, we need to understand the following major issues:

- What is the nature of the genomic data?
- What hypotheses are we testing in racial genetics?
- What methods are used for testing hypotheses?
- Are the data being treated properly in testing those hypotheses?
- What would the significance of a test look like?

These are all heady questions that are sometimes ignored because of the complexity of modern genomic data. Indeed, it is true to say that researchers have rarely been faced with such complex data, or with data that have such enormous potential for information – or misinformation. This is because the data involved are multidimensional, taking us into a province that is difficult for the human mind to perceive, let alone to decipher. To approach such complex data, we need to use shortcuts; and, because we are a very visual species, scientists have developed several data-reduction techniques that reduce the complex data sets to visually appealing results. This chapter addresses the basic issues that are involved in the application of genome-scale information to racial science, and we start by discussing the nature of the genomic data and how they are ascertained, followed by detail on the various data analysis methods that have been developed. We look at hypothesis formulation and testing in Chapter 7.

Data Set Ascertainment

Establishing which data are to be used in a human genome comparison study is critical to its final outcome. Different methods exist for setting up the data set (ascertainment) in the first place, so while two researchers might be confronted with the same raw DNA sequence data for a bunch of individuals, they might end up with rather different primary data sets for making inferences. Because the choice of markers may sometimes be biased by the researcher's initial ideas about the structure of the data set, problems with ascertainment bias might enter into consideration.

At one end of the ascertainment bias spectrum in human genome comparative studies lies an approach called ancestry informative markers (AIMs). In this ascertainment strategy, only those markers that are informative for identifying a researcher-designated grouping are retained in a data set, and all the rest of the data are excluded. An AIMs approach starts with a small subset of individuals whose genomes are compared; all positions in the genome that are diagnostic of the subset, or nearly so, are designated AIMs. Consider a ride on a subway car in New York City. All of the individuals in the car have their genomes sequenced and the results are scanned for polymorphisms that are the same in all individuals in the car but different from other humans. These polymorphisms are AIMs for the people in the subway car. This approach has an extreme ascertainment bias toward the researcher's original notion of what groups might be represented among the individuals in a data set. In other words, the grouping of humans in a subway car in New York City may not be the best way to stratify the genomes in a study.

Somewhere in the middle of the ascertainment bias spectrum lie approaches that use a small subset of individuals to establish positions in the genome that are polymorphic and thus potentially informative about the relationships by descent of the individuals within the study. This ascertainment approach is commonly used to establish "human genome variation panels" that are used to subsample the genome for variation. And it makes sense to the extent that it eliminates all sites in the genome that are ancestrally polymorphic, or that are absolutely invariant among all humans. These panels typically turn up anything from half a million to over one million sites in the human genome that can be used to make inferences about relatedness among the individuals in the study. At the extreme opposite end of the bias spectrum to AIMs are approaches that retain all of the sequence data in the genomes under study. They are rarely used because of the complexities of comparing billions of nucleotide positions across thousands of individuals.

Most race-based studies of human genomes attempt to reduce the complexity of genomic data using visual methods that involve clustering in one way or another. Some of the approaches developed for human diversity studies are more rigorous statistically than these, but turn out to be more useful in developing hypotheses than in testing them.

Treeing

Branching diagrams provide a way for researchers to visualize the large nucleotide data sets available to them for human variation studies. Recall that early analyses of human variation via blood groups and protein electrophoresis often used trees to summarize and display data. We addressed the nature of those trees in Chapter 4, and only need to remind the reader here that trees are branching diagrams in which the nodes can, under some conditions, represent common ancestors. As we saw, branching diagrams can be constructed from genomic data using similarity (via data-reduction techniques), parsimony, or likelihood. When similarity is used, the complex sequence data are reduced to quantitative numerical information that can be used to construct a dendrogram (a graph that summarizes the similarities). While many researchers use such diagrams as phylogenetic trees, others argue that such diagrams do not truly represent data-driven ancestor–descendant relationships. Given a matrix of sequences and a tree topology, every node in the tree can be reconstructed. This might seem like nitpicking, but it is the ancestors in a tree that make the approach "phylogenetic."

The most powerful aspect of a phylogenetic tree is that its terminals will form groups based on common ancestry. In these cases, groups of individuals can be classified as "polyphyletic," "paraphyletic," or "monophyletic." Monophyletic groups (those groups with a single common ancestor to the exclusion of all other individuals in the analysis) are the only meaningful ones if you wish to assess whether all individuals in a group under study are members of the same species (Figure 6.1). In other words, monophyly is the

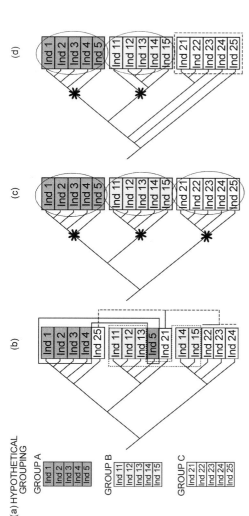

Figure 6.1 Phylogenetic tree with individuals as terminals showing hypothetical grouping (a) and three tree shapes where the hypothetical groups listed in (a) are polyphyletic (b) and monophyletic (c). Part (d) represents a complex situation in which some of the hypothetical groups are monophyletic, and one is paraphyletic. Note that in (b) the hypothetical groups are scattered throughout the tree, and that none of the three hypothetical groups in (a) can be determined to be monophyletic. The dotted lines circumscribe the hypothetical Group A, the solid lines circumscribe the individuals in hypothetical Group B, and the dashed lines circumscribe the individuals in hypothetical Group C. In (c) that all three hypothetical groups are monophyletic, and the common ancestors of the three groups are indicated by stars. In (d) hypothetical Group A and hypothetical Group B are monophyletic, and the stars again indicate the common ancestors of these two hypothetical groups. Hypothetical Group C is not monophyletic, but is rather paraphyletic. The paraphyly of this group is caused by the placement of individuals 24 and 25, which are not in Group C to the exclusion of other groups.

arbiter in tests of group inclusion. As Figure 6.1(c) shows, monophyly visually supports the grouping of the three hypothesized groups in Figure 6.1(a). Even more important is that the common ancestor indicated by stars in Figure 6.1 can be determined, along with the diagnostics (i.e., the sequence positions that are uniquely shared among the descendants of the common ancestor).

Tree-based approaches were used early on in attempts to understand human variation (remember mitochondrial Eve and Y chromosomal Adam). Due to the computational intensity of the approach, and the lower computational demands of methods that use data reduction (discussed below), this tactic is currently used sparingly. When it is used, however, any test of a hypothesis of grouping along racial lines can be rejected. For example, one of us took the 1000 Genomes Project (1000GP) data sets and examined various hypotheses of grouping using monophyly as the testing criterion. Enforcing this criterion means that all individuals in a hypothetical group (all people from Asia, say) would have the same single common ancestor, to the exclusion of all other people in the study. No single subpartition of the data set (either by X chromosome, mtDNA, Y chromosome, or chromosome 21) was found to support any racial grouping, as monophyly of the individuals expected to be in particular groups turned out to be nonexistent. Additionally, when all the data were combined the picture became even hazier with respect to monophyly, and more problematic in terms of racial grouping. This result is expected, because as more and more parts of the recombining genome are added, admixture increasingly becomes a property of the data set, and monophyly erodes.

Clustering

There are several clustering approaches for analyzing genomics data. All of them (even the tree-clustering methods discussed above) rely on data-reduction techniques and the visual presentation of their results.

Pie diagramming: Perhaps the simplest approaches are the ones that produce pie diagrams. This approach was used extensively by Cavalli-Sforza and colleagues in analyzing the pre-genomic data in their tome *The History and Geography of Human Genes*, and some current researchers use pie diagrams to summarize their sequence information. But there are several ways in which these pie diagrams can be drawn, depending on the ascertainment procedure used to construct the data matrix. If AIMs are used, then the pie diagram will represent just those sites in the genome that the researcher had determined were informative for a race-based hypothesis. If the entire genome is used, then the pie diagram will have a large representation of invariant sites (Figure 6.2).

With this caveat, we suggest that all the pie diagram analyses that we are aware of lack diagnostic value for human genome comparison studies. Some researchers will shift the significance of sites from being "diagnostic" to "private," which simply means that a site is found exclusively at a particular locality but is not diagnostic. Figure 6.3 shows such an analysis of genomes from the LWK population in the 1000GP. Note that about 75 percent of the pie in this diagram is either dark gray (found on all continents) or light gray (found

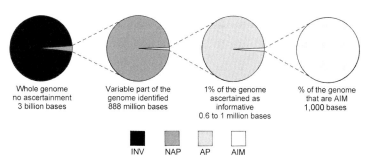

Whole genome	Variable part of the	1% of the genome	% of the genome
no ascertainment	genome identified	ascertained as	that are AIM
3 billion bases	888 million bases	informative	1,000 bases
		0.6 to 1 million bases	

INV   NAP   AP   AIM

Figure 6.2   Pie diagrams of a hypothetical analysis of typical human genomes. The genome can be partitioned into an invariant portion (INV); that portion of the genome that is variable but not ascertained (NAP); that portion of the genome that is variable and ascertained (AP); and that portion of the genome that is ancestry informative markers (AIM). The first pie diagram represents the entire genome (INV + NAP + AP + AIM); note that the majority of the pie diagram is INV. The second pie diagram represents that portion of the genome that is variable (NAP + AP + AIM); note that the majority of this pie diagram is variable but not ascertained (NAP). The third diagram represents that part of the genome that can be ascertained for human genome-level studies (AP + AIM); note that the majority of this pie diagram is single nucleotide polymorphisms (SNPs) that are ascertained but not AIMs. The fourth diagram represents that part of the genome that are AIMs (there are on the order of 1,000 AIMs in human genome studies).

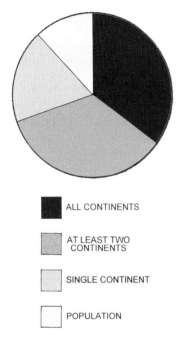

ALL CONTINENTS

AT LEAST TWO CONTINENTS

SINGLE CONTINENT

POPULATION

Figure 6.3   Typical pie diagram for a particular geographic region (the LWK Luhya, Kenya population; for more information on this specific population see Figure 4.3) from the 1000GP showing the proportion of DNA variants (SNPs) in the LWK population that are found on all continents, at least two continents, private to a single continent (Africa), and private to a population (population here is defined as the LWK population).

on at least two continents). Pie diagrams for other populations are similar, with ranges of dark and light gray regions from 75 to 90 percent. These levels of geographic spread indicate to us widespread commonality of sequence identity across all the 2,000 or so individuals in the 1000GP. The remaining sites are split between those that are private to a continent or private to a pre-described population. Both contribute to the sites that researchers use to make inferences about race, but both categories lack fixed and different sites between continents, emphasizing the lack of diagnosability using these variants.

Structuring

One way in which we can examine genetic variation is by looking at how it is distributed among the individuals in a data set. To accomplish this, researchers have tried estimating the probability that each individual in a study belongs to a particular ancestral population (Figure 6.4). A group of population geneticists developed an incredibly important computer program called STRUCTURE to do just that, and to date it has been cited in over 35,000 research papers (that's probably three times the number of citations the average researcher will accrue in an entire career). Jonathan Pritchard,

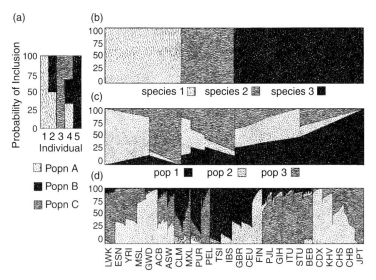

Figure 6.4 Examples of STRUCTURE analyses at the species level, population level, and geographic level.

Mathew Stephens, and Peter Donnelly actually developed the program to analyze genetic variational structure in nonhuman natural populations, but because they were also interested in human genetics it began to be used for humans as well. STRUCTURE is mostly useful at reducing the visual complexity of genomic data sets.

The visual outcome of a STRUCTURE analysis is a series of stripes that indicate the species or population (or whatever entity can be justified) of origin of the individuals in a study (Figure 6.4). This visual construct is highly dependent on the preset number of populations (loosely, the K parameter used in the model developed for STRUCTURE). Researchers can preset the numbers over a range of values, and then assess the fit of the data to the model with the various K parameters. This will yield the K value with the highest probability.

As for the applicability of such approaches to humans, in his treatment of race the journalist Nicholas Wade focused on a 2002 study by Noah Rosenberg and colleagues. This group used 377 genetic markers, chosen to maximize the potential for discovering variability, on 1,056 individuals from 52 discrete populations. In other words, they used a severely ascertainment-biased data set. Their original STRUCTURE analysis is shown in Figure 6.4, which suggests that there are six populations based on the statistics of the various K values examined. And indeed, to the naked and undiscerning eye the shades appear to cluster in six groups. Five of the six populations coincide with geographic groupings of people, and this is what Wade focused on in his conclusions about genetics and race. But now look at the STRUCTURE diagram with our hypothesis in mind, and with a discerning eye. If the criterion for existence of a race is a visual clustering of shades at $K = 6$, then we have races. But there is no such definition. Frankly, we find it difficult to conclude that such analyses say anything other than that there is some underlying structure in the data (we have an idea what it might be, and we will discuss the nature of the visual appeal of STRUCTURE in Chapter 7). But races? There are definitely no species here, as there is no diagnosis. But if we want to call the six colors in Rosenberg et al.'s figure subspecies, then we are still in a bind. Calling these groupings subspecies is simply a hypothesis of a group's existence, and we know we can easily reject that on the basis of two bits of information. First, as we pointed out, there is no diagnosis. Second, we know that individuals from the six proposed subspecies interbreed, and that they do so at high frequency.

Wade also cites Rosenberg and colleagues' 2005 paper in which they slightly reduced the sample size (down to 1,048 from 1,056), but this time with almost 1,000 genetic markers. The results are pretty much the same as with the 2002 data set, with an identical $K = 6$ being the value for the number of entities. Let us simply point out here that the same sample size problem existed as before, and the ascertainment bias was still extreme. Even more significantly, piling on more loci that were ascertained in the same way as the first 377 markers failed to strengthen the inference at all. In a third study that has often been cited as proof of the existence of six or seven "races" of humans, Jun Zi Li and colleagues in 2008 upped the ante of loci several orders of magnitude and examined over 600,000 SNPs. Not surprisingly, they obtained $K = 7$ (one louder than $K = 6$) as the statistically significant indicator of population structure.

Since 2008, several papers have used the STRUCTURE, ADMIXTURE, and fineSTRUCTURE routines, among others, to address the huge quantity of human genome data accumulated over the last few years. We have neither the time nor the patience to look at them here. With that said, we can point out that studies using STRUCTURE or STRUCTURE-like approaches to human population structure over the last five years fall into one of two kinds. The first are fine-grained studies that look at localized populations like Ashkenazi Jews, British populations, Tibetan highlanders, and sub-Saharan African populations. These studies are basically designed to understand the ancestry of localized groups of people, and they commonly use membership in a cultural group as a stratification approach. Studies of the second kind take a global approach to human population structure, and are dyed-in-the-wool race-based. Both categories of human population structure studies have had to deal with expanding population sizes, going from the 1,000–2,000 individuals in the 1000GP and the Human Genome Diversity Project (HGDP) to literally millions of individuals. As you might imagine, computation time for these larger data sets is a problem. Several tools have been developed to handle this issue. Prem Gopalan, Wei Hao, David Blei, and John Storey call this tera-sampling, and have developed a tool called teraSTRUCTURE to handle the tera-data. We describe here the major result from their analysis of 1,718 people at 1,854,622 SNPs. As the number of people included in the STRUCTURE analysis went from 1,000 in

2008 to 1,700 in 2015, the STRUCTURE analyses witnessed more and more of the blocks that had been solid-colored in 2008 beginning to be bleached with colors from other population blocks. Imagine a picture that was only a little out of focus in 2002 getting blurrier in 2008, and then losing even more definition in 2015.

Sure, in these publications the structure diagrams are pretty. But it is something else entirely for them to be indicators of racial separation. It is also significant that, as Sarah Tishkoff and colleagues have shown, optimal K (number of potential divisions) leaps from 7 to 13 after the addition of several African genomes. Adding even more genomes will more than likely bleach out much of the appearance of divisions in these STRUCTURE analyses. This bleaching makes sense, as the more humans are added to a study, the more likely it is that admixture will be detected, leading to bleaching of boundaries.

Another more serious problem with interpreting STRUCTURE analyses too literally is summed up in John Novembre and Benjamin Peter's statement about the siren visual appeal of these approaches: "A final precaution, and one of broader societal relevance, is that a viewer can become misled about the depth of population structure when casually inspecting visualizations using methods such as PCA, ADMIXTURE, or fineSTRUCTURE." A reminder that the eye can indeed be tricked.

PCAing

Another method of clustering individuals in genomic studies is to use the principal component analysis (PCA) approach, or one of its derivatives. Olivier Francois and colleagues have reviewed how the PCA approach was integrated into human population genetics. Their description of PCA in population genetic studies is interesting but technical, and might help clarify how the approach is used in human population genetics. If the quote doesn't help, don't worry, we have tried to paraphrase it below:

> One way to explain PCA is as an algorithm that iteratively searches for orthogonal axes, described as linear combinations of multivariate observations, along which projected objects show the highest variance, and then returns the positions of objects along those axes (the principal

components [PCs]). For many data sets, the relative position of these objects (e.g., individuals) along the first few PCs provides a reasonable approximation of the covariance pattern among individuals in the larger data set. As a result, the first few PC values are often used to explore the structure of variation in the sample.

Principal component analysis is a data-reduction approach that is used to simplify the appearance of complex data sets (genomic data sets are incredibly rich and complex with respect to information). All data sets have variation in them, and it is this variation that makes the data set informative. Data reduction is accomplished by discovering which variables in the data set contribute the most to the structure. This step is usually accomplished by determining which variables best explain the variation in the total data set. Let's look at a simple example. For a one-dimensional PCA with two variables, the data are examined and the variable contributing the most to the structure of the data is retained while the other is dropped. The retained data are rescaled for graphing purposes, and a line graph is produced. The position of the data points on the linear scale indicates the similarity of the data points. For a two-dimensional PCA with $N > 2$ variables, the top two variables (or components) are retained (actually, any two can be retained, but most commonly it's the top two). These variables are rescaled and graphed, and their position in two-dimensional space (with X and Y axes) tells the researcher about the relationship of the data points (which in a genetic study will represent individuals). For a three-dimensional PCA with $N > 3$ variables, the top three variables (again, it can be any three variables, but the most common way of doing a three-dimensional PCA is to use the top three variables) are chosen and graphed in three dimensions (with X, Y, and Z axes). The positions of the variables in the three-dimensional space indicate their relatedness to each other.

As we pointed out in Chapter 5, Cavalli-Sforza and his colleagues used the PCA approach in their seminal and widely influential work in the 1990s on the history of human genes. To many researchers who use the approach, PCA seems like the perfect technique to reduce noise in the face of the complex data of the genome. In 2019 we reviewed the literature for the use of PCA in human population genomics studies and found at least 33 large-scale genomic studies that use PCA published in the last five years. Table 6.1 summarizes several of these studies done on worldwide analyses of genomic data,

Lead author	N	Pops	SNPs	% PC1/PC2	Clusters	Note
Li[1]	**938**	**51**	*	**86**	**4 or 5**	**F_{ST} of 51 × 51 populations**
Paschou[2]	255	11	10,805	45	4	**30 PCA-correlated SNPs**
Auton[3]	3,845	11	443,434	7	7 or 8	Clusters overlap
Nassir[4]	**1,620**	**20**	**93 AIMs**	**80**	**6 or 7**	**93 SNP AIMs**
Sudmant[5]	2,504	26	68,818	8	3 or 4	Africa Asia out, others overlap
Biswas[6]	1,043	52	46,000	<10	6 or 7	Using SNP correlated to PC
Lek[7]	60,706	5	5,400 c/e	NA	5	5,400 common exome SNPs
Martin[8]	2,504	26	Filtered	14	3	Extremely admixed
Cherni[9]	**2,984**	**57**	**399**	**87**	**5–9**	**399 hand-picked SNPs**

*This study is unique in that it computes the F_{ST} (a measure of differentiation) based on 650,000 SNPs to use in the PCA, which reduces the data set considerably. [1] Li et al. (2008). Worldwide human relationships inferred from genome-wide patterns of variation. [2] Paschou et al. (2007). PCA-correlated SNPs for structure identification in worldwide human populations. [3] Auton et al. (2009). Global distribution of genomic diversity underscores rich complex history of continental human populations. [4] Nassir et al. (2009). An ancestry informative marker set for determining continental origin: Validation and extension using human genome diversity panels. [5] Sudmant et al. (2015). Global diversity, population stratification, and selection of human copy-number variation. [6] Biswas et al. (2009). Genome-wide insights into the patterns and determinants of fine-scale population structure in humans. [7] Lek et al. (2016). Analysis of protein-coding genetic variation in 60,706 humans. [8] Martin et al. (2017). Human demographic history impacts genetic risk prediction across diverse populations. [9] Cherni et al. (2016). Genetic variation in Tunisia in the context of human diversity worldwide.

Table 6.1 Summary of genome studies that use worldwide sampling and PCA. Reproduced with permission from DeSalle and Tattersall (2019). Bold entries are more subjective in their ascertainment of markers.

showing that there is some degree of variation in the number of clusters discovered using PCA for human populations. Depending on how the PCA is constrained, this approach will give anywhere from three to nine global clusters. The analyses are dependent on sample size (N) and on the number of populations used in the analysis.

Clustering is evident in most of these studies, but it is difficult to make a consistent interpretation of the data overall with respect to the reality of geographic or racial clustering. This is because of the subjectivity, or bias, with which some of the studies ascertain their data (bold entries in Table 6.1). When there is a higher level of objectivity (nonbold entries in Table 6.1), far less of the variance in the data set can be explained. While not quantifiable, there is a discernible decrease in discrete clustering when higher objectivity is used to ascertain SNPs in such studies.

David Reich and colleagues have developed an improvement to the PCA approach, called EIGENSTRAT. This approach established a statistical significance for the principal components that are generated in PCA. The name EIGENSTRAT comes from the use of mathematical eigenvectors to stratify (group) individuals in genome-wide association studies in a statistical framework. The statistical approach might allow a researcher to have more confidence in the existence of a cluster in a PCA analysis. One of the important applications of this approach is that the various principal components that are statistically significant for a data set can then be used as markers for the structure of the data set. But is this a reasonable way to view PCA analyses, even with statistical inference?

To answer this question, let's look at what PCA analyses tell us about differentiation within a data set from a statistician's viewpoint. Here we turn to one of the main researchers in PCA and clustering-based analysis, who focuses on nongenetic research: Anil K. Jain. His perspective is one of bare-boned science. Of clustering, he says: "Organizing data into sensible groupings arises naturally in many scientific fields. It is, therefore, not surprising to see the continued popularity of data clustering. It is important to remember that cluster analysis is an exploratory tool; the output of clustering algorithms only suggest hypotheses."

IBDing

A novel clustering approach is based on estimates of identity by descent (IBD) of individuals in human genetic variation studies. The IBD concept is an old one in human population genetics, and it makes sense as an indicator of descent and relatedness. It has been used by Eunjung Han and colleagues to decipher the complexity of genomic information for over 700,000 people living in America (Figure 6.5). This study, a collaboration of several academic institutions and the direct-to-consumer company Ancestry.com, used a modern conceptualization of IBD that involves long stretches of the genome that are identical by descent between two people – that is, that have been inherited from the same common ancestor – to generate similarity measures for various types of cluster analysis. They also used the massive metadata (over 20 million records) that direct-to-consumer companies collect, along with the sequence data, to establish relatedness via IBD. To show how complex this data set is, for 700,000 individuals there are 2.5×10^{11} pairwise IBD comparisons. Add to this that over 700,000 polymorphic genomic positions were examined, and the unimaginable vastness of the data set should be evident. The authors of the paper were able to whittle the number of "individuals" in the study down to 5×10^8 IBD measurements by clumping closely related individuals together and thus reducing the number of genomes analyzed. Once the IBD measurements are made, they can be used to hierarchically cluster the "individuals" in the study. Further, a set of variables can be computed for each of the 48 mainland states that are amenable to analysis, using PCA approaches, to give a two-dimensional graph of the 48 states in PCA space (Figure 6.5).

Interestingly, when they used a PCA clustering method, states from distinct geographic regions (Midwest, East, West, and South) clustered in different parts of the two-dimensional graph. In other words, by stretching a map of the mainland United States the authors could neatly accommodate the position of each state in the two-dimensional graph to it. While their results demonstrate a correlation between historical hierarchical clustering and geographic location in the United States, the authors are very cautious in making historical inferences. They took a minimalist racial approach to their data analysis, eschewing racial or ethnic designations during the analysis part of their

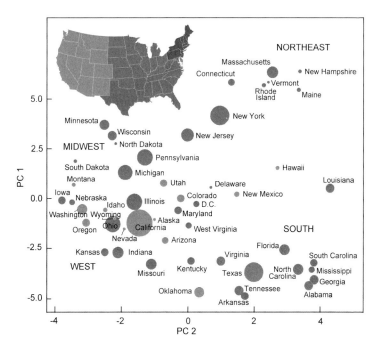

Figure 6.5   Using IBD to construct a PCA rendering of different US populations and their relatedness. The two-dimensional distribution of the states resembles their geographic distribution.

study, and only bringing ethnic background into the picture to interpret the complexity of the data set. They offer the reasonable and correct caveats that, first, while the clusters appear in their plots this does not mean that there isn't a lot of interbreeding between the clusters; and second, their data set, while showing some structure, also demonstrates a great deal of heterogeneity in the genomes of people living in the United States.

But what about race? The authors are very clear that this method works better the more cohesive or closely related the individuals in the analysis are. For instance, the authors could make some very fine-scale inferences about the demographics and migration of historical French-Canadian genomes south into New England and Louisiana. So, the approach is very amenable to understanding the demographics and history of recent, very discrete migration events. But on larger scales its power is more questionable. We have already discussed the caveats involved in scaling-up, and if the trend continues with larger and more inclusive grouping, its value as a demographic tool will diminish. We suggest that, for races, the approach will show exactly what has happened, namely massive amounts of interbreeding and genetic contact among people across the globe.

The Bottom Line on -inging

Let's summarize the approaches we have discussed here and their relevance to racial science in the context of genomics. After all, much of the recent misunderstanding about the reality of race is based on some of the studies we have discussed:

- First and foremost, the most rigorous and exacting approaches to testing hypotheses about biological races – using trees – reveals a lack of monophyly at all levels, and for all partitions of the genome. This makes sense because of the extensive amount of interbreeding of people on our planet.
- Next, visual approaches like pie diagrams cannot go very far in testing a hypothesis about biological race existence. While they seem to be visually leaning toward differentiation, they really are not; and, in addition, we have to be careful about what is in the pie diagram. Many pie diagrams will only include a small proportion of the overall genome, or even of the variable part of the genome. Pie diagrams for different populations or groups of humans can be compared and will show large differences for small portions of the genome; but when the pie diagrams are based on large portions of the genome, they are quite similar when compared across groups of people.
- Third, while STRUCTURE analysis can also give us a good, colorful view of the potential differentiation of people on the planet, such analyses do not test hypotheses and thus are not telling us something about the scientific validity of the race hypothesis.

- The recent use of IBD clustering, while giving exquisite pictures of fine-grained differentiation, will not be so discerning when testing broader hypotheses about biological race.
- Finally, hierarchical clustering methods need to be considered with the utmost care when it comes to making statements about biological race.

We suggest that the best way to look at the results of any clustering analysis is to assume not that they test hypotheses, but that they are extremely useful in visualizing data, and in helping to formulate strong hypotheses. We return to Anil Jain's statement about clustering methods – "the output of clustering algorithms only suggests hypotheses." We suggest, therefore, that PCA clustering "evidence" for differentiated human populations is not an approach to cite as proof of racial differentiation in our species. To the credit of several of the researchers who do use PCA and other clustering approaches, they often point this out as well. As Novembre and Peter note: "untrained eyes may overinterpret population clusters in a PCA plot as a signature of deep, absolute levels of differentiation with relevance for phenotypic differentiation." The visual appeal of clustering can be considered the siren song of racial research; and if they want to prevent the ship of science from crashing onto the rocks, researchers will need to understand some of the basic underpinnings of how we humans attempt to understand the world around us. So, while we have examined many of the results that have led some to inaccurately claim a genetic biological basis for race, we now need also to examine some of the philosophical aspects of racial science.

7 Race in Medicine and Complex Phenotypic Studies

Why do some researchers care so much about race in their experimental designs? It is easy to understand why a racist forced by inherent bias would take this approach. But why would a nonracist biologist insist on doing racial science? As far as we can tell, there are two major reasons for this: medical expediency and the discovery of the genetic basis of complex traits. We would suggest that the first of these is a red herring, and the second a brick wall.

Some Background

Both red herring (race in medicine) and brick wall (race in complex phenotypes) use approaches that at some phase of the study involve "stratification" of the subjects, by race or ethnicity. The genetic basis of diseases and their epidemiological aspects are, of course, important endeavors in modern medical science; and it is safe to say that the public appeal of such initiatives as the Human Genome Project derives from the promise of medical advances. That's because, if we have a genetic basis for a disease we will also have an etiological basis for it, and methods for early detection and treatment can be developed. To scientists who use this approach, race is potentially important in the discovery, development, and application of cures and treatments for human ailments, and the methods used in genome-wide association studies (GWAS) provide an obvious starting point. Genome-wide association studies are conducted by collecting DNA sequence or SNP (single nucleotide polymorphism) data for a large number of individuals who have a trait or a disease of interest (called "cases"), along with people who don't have the trait or disease ("controls"). The sequences are scanned by computer algorithms that can detect SNPs associated with the disorder or trait. Such SNPs are usually in or near genes that are the basis for the trait or disease. These studies are computationally intense, and stratifying the data along racial lines can often speed up the discovery of associated SNPs. As we stated in our book *Race: Debunking a Scientific Myth*:

> To be clear, the major advantages of using race in GWAS are that stratifying data along racial lines makes the techniques rapid, simplifies the collection of baseline information, and yields enhanced statistical significance. All three of these advantages are, of course, desirable in studying disease.

But we suggest that other approaches that avoid racial stratification are more appropriate. These include cohort analysis and trio analysis; these are preferable because while they require more effort, they also make fewer assumptions and use the principles of classical genetic analysis more appropriately. They also take a more sophisticated view of the dynamics of genetic disorders. Genome-wide association studies require that researchers take advantage of stratification as a quality control of the data being analyzed, because any data that are deemed to be biased are either removed or controlled for. And it is far from clear that this is the case when stratification by race is used. Another more insidious problem with GWAS case–control design is that, for availability reasons, it has been easier to conduct such studies on people of European descent, leaving other ethnic groups out in the cold. In 2009 Anna Need and David Goldstein pointed out that, of the 370 or so GWAS studies conducted up to that time, 85 percent were on European populations. A decade later, Giorgio Sirugo, Scott M. Williams, and Sarah A. Tishkoff surveyed the ethnicity of subjects of GWAS subjects up to that date, showing that the percentage of European individuals in the GWAS catalog (the database for GWAS studies) had only dropped to 78.4 percent, with 10 percent of the individuals in the catalog being Asian, and only 2 percent African. The bias in ethnic focus has not improved in over a decade, and we doubt it ever will.

The Red Herring

Entire books have been written on the inadequacy of race-based medicine, but sadly the use of race in medicine and in pharmaceutical development persists. At a very basic level, there are two ways to use race in these contexts. The first concerns using race to stratify a data set to extract information about suscepti-bility to a disorder in some (racial or ethnic) group. This approach looks for the genetic basis of a disorder that can then be leveraged into some pharmaceutical or gene therapy development. The second approach is for nongenetic purposes, and involves using racial stratification to exam-ine environmental factors involved in a disease. We often forget or ignore the fact that a phenotype is the product both of the genes involved *and* of the environment, and the social and cultural contexts of disorders are often more important than any genetic component. An example is given by the recent COVID-19 pandemic: While the disease raged, researchers discovered that minority populations were disproportionately impacted by the virus causing it. This discovery was based on racial science, but while it proved to be entirely helpful and useful in treating COVID-19 patients, that was not because the genes of different ethnic groups governed susceptibility to the disease. Instead, differences in sociocultural factors were hugely significant.

Some researchers suggest that using ethnicity in GWAS and other kinds of disease gene discovery studies allows for a better understanding of the genetic architecture of the disease in ethnically distinct subgroups or populations. After all, numerous disorders are purportedly found in higher frequencies in certain ethnic groups. Both Vivian Tam and colleagues and Huaying Ling and colleagues have made this claim with respect to ethnicity-tinged GWAS studies. However, such claims can mean many things with respect to the genetic basis of disease. The genetic architecture of a disorder, or of a disease, also includes the role of the environment in its expression. We suggest that sorting out the environmental component of a genetic architec-ture is the major reason to apply ethnicity to these kinds of studies; and in the end, it will be its strongest component, vastly outweighing any single gene or locus.

Much of the literature on race and medicine involves a defensive approach that starts with an admirable rationale, namely that researchers wish to use race to uncover the biology of a disorder for the development of therapeutic treatments. But we would like to turn this rationale on its head, using Jonathan Kahn's suggestion that "medical researchers may say they are using race as a surrogate to target biology in drug development, but corporations are using biology as a surrogate to target race in drug marketing." In other words, we think that anyone using racial and ethnic stratification in science needs to show that it is not for the purposes of marketing or some other nefarious objective. While we realize that this approach is rather dark and laden with suspicion, we think it is better to make sure of this rather than simply to assume that race is adding anything to the science of disease genetic architecture. To strengthen our position, we refer to the well-known and often-cited BiDil case. Why use a worn-and-torn example like BiDil? Because the drug is still in use, the dynamics of its employment and misuse are still being dissected, and it remains a classic example of race-based treatments. It serves as the perfect paradigm for this approach to medicine.

BiDil was developed as an anti–heart attack agent by the company Nitromed, which claimed that it had superior efficacy among African Americans. When it was in development it could be justified as a race-based medicine basically because the available technology could not identify individuals with the genetic profiles that would best respond to the drug. Two important facts that were not specified clearly by the marketers are that, first, not all African Americans respond to the drug, and second, that some people of European descent respond positively to it. As Dorothy Roberts has suggested:

> Portraying BiDil as a solution to a racial gap in mortality implies the gap stems from racial differences in disease and drug response. Adding a genetic explanation for this difference attributes health disparities to flaws inside black people's bodies rather than to flaws in the society they live in. It supports the increasingly popular but misguided view that the tiny per-centage of genetic difference among human beings is distributed by race and that this difference creates inequities in health.

Indeed, the misguided view that Roberts mentions is not only prevalent in describing inequities in health, but also in describing inequities in other complex traits that we think are important to human existence. Like Dorothy Roberts, we are not making these claims to deprive a specific ethnic group of life-saving treatments. Indeed, completely pulling BiDil from the market would be a counter-productive act to human health in general. Roberts summarizes this nicely:

> It should be made more widely available—without regard to race. We simply see no justification for marketing medicines according to race and worry about their potential to divert attention away from more significant social reasons for health disparities. Studying and eliminating the social determinants of health inequities is a far more promising course than searching for race-specific genetic differences.

There are two simple solutions to this problem. The first solution that Roberts emphasizes is to create a more just society ("A more just society would be a healthier one"). The second is to make pharmaceuticals and other therapies available based on individualized genomics. Here, Roberts recognizes that the individual's genetic profile or the environmental component affecting the trait or disease is more important than his or her race. Researchers used the racial approach when BiDil was first developed because of technical limitations in sequencing human genomes. Those limitations no longer exist when we have the capacity to sequence disease-related loci of humans very inexpensively.

The Brick Wall

The brick wall to which we refer is the use of the ethnic or racial stratification (grouping) of people to understand the genetic architecture of complex traits, or phenotypes. To illustrate the importance of this subject, let us again refer briefly to Nicholas Wade's book *Troublesome Inheritance*, more fully discussed in Chapter 9. The first part introduces Wade's view of the genetic basis of racial differentiation; because he felt he had clearly proven that races do exist, he felt compelled to continue in the book's second half to describe certain human phenotypes that are racially connected. It is essential for the proper understanding of race that we closely examine the genetic architecture of the complex phenotypes that are discussed by Wade and other proponents of racialized views of complex human behaviors and social "traits."

The endeavor of associating complex human traits with particular genes is not a new one. In 1975, Harvard University Press published *Sociobiology: The*

New Synthesis, a magisterial survey of behavior across all groups of social animals by the myrmecologist E. O. Wilson, whose accomplishments as an entomologist and biodiversity expert specializing in the highly social ants are unassailable. Organized into three main sections, his book outlined a research program for studying animal behavior. It started with the principles of evolutionary analysis, went into the mechanisms of social interactions, and ended with a comprehensive survey of social organisms in the natural world. In our minds, and those of many others, the book's only problem was its chapter 27, which suggested that knowledge of behavior in other animals could advance our understanding of similar behaviors in humans. Wilson went so far as to claim that "the systematic study of the biological basis of all social behavior" could be carried out using his comparative approach. The problem started with his claim that "all" social behavior could be studied this way. Perhaps because of the pointed negative response to the book, the study of sociobiology was renamed "evolutionary psychology," under which guise the field has persisted into the present. But while we applaud the comparative method, we seriously doubt that the complex human behaviors involved in religious attitudes, political preferences, sexual behaviors, aggression (warfare), and, most of all, intelligence, can be dissected and understood as genetically distinct entities. It just ain't so.

Here we will quickly look at the well-worn subject of race and intelligence and other cognitive expressions. In his recent book *Two Americas*, Charles Murray, co-author of the much-discussed 1990s bestseller *The Bell Curve*, returned to the proposition that America's current social ills can be traced to the unequal intelligence capacities of different ethnic groups. We return to this proposition in Chapter 9, but meanwhile we need to point out that behavioral traits such as intelligence are even messier with respect to discrete explanations than physical ones. Two approaches have typically been taken to understanding the genetic architecture of complex behaviors: the candidate-gene method and GWAS. Both approaches have (necessarily) shown some associations; but again, both approaches suffer from the problems we discussed earlier with respect to disease genes. As an example, consider that, in a large study of the political attitudes of Australian citizens, Peter Hatemi and colleagues found that large numbers of genes involved in neural processing (production of serotonin, glutamate, dopamine, and so forth) were also associated with beliefs connected to both liberal and conservative political attitudes. In addition to the fact that the associations discovered in the study were difficult to replicate, no single associated gene or gene region could account for more than 12 percent of the variance in the population. This means not only that the genetic influence on political leanings is minimal, but that it is better considered as a complex trait distributed along a bell curve, rather than as the specifier of the discrete states that the study used to come to its conclusions.

Intelligence is probably the most visible, important, and hotly contested complex behavior that humans exhibit. Let us first point out that humans are unique in the ways in which they use information to cope with their environments. There is no question that humans are highly variable in intelligence – which is, indeed, what has made this so-called trait the poster child of racial science, and eventually of racism. Let us look here at a recent example of well-intended racial science gone wrong. A strength of this study is that it used millions of subjects' genomes (partially provided by a direct-to-consumer [DTC] genome sequencing company, 23andMe). The DTC companies don't just sequence genomes for customers; they also collect a considerable amount of what are called metadata, or information about the customer. The metadata are sometimes broad in context, but one aspect of the 23andMe metadata was the inclusion of self-rated educational attainment. This gave the researchers using the genomics/metadata the opportunity to examine genomic associations with such attainment. As a necessary caveat, we should note that we are acutely aware that educational achievement is not necessarily a good direct proxy for whatever it is that we might want to call "intelligence." But it does nevertheless tell us something about the architecture of intelligence as humans develop. And just as notable as this study's scholarly content is the response it has received from educators and journalists.

The study revealed that 1,271 genes are involved in only 10 percent of the variation in educational attainment. Suffice it to say, this is a lot of genes to explain rather little of the variance in any behavioral capacity, let alone to serve as a proxy for intelligence. The data suggest an inherent complexity in the trait, and it gives us some idea of the difficulties we face when merely trying to define "intelligence." If all the folks who read the paper and embraced its findings had realized this complexity and those limitations, then all would have been well. However, since educational attainment has become a racialized issue, some well-respected educators and not-so-well-respected commentators embraced the study to suggest that we now have a significant basis for understanding the racial component of educational attainment. In a *New York Times* editorial, Kathryn Paige Harden, a psychologist whose specialty is adolescent development, suggests that there are two ways we can use this genetic "information." The first is to understand that success in our educational system is not entirely reliant on merit. As she says, we don't earn our genes; and because the inheritance of our genes is somewhat random, then "everyone should share in our national prosperity, regardless of which genetic variants he or she happens to inherit." Second, she suggests that knowing the genetic architecture of educational attainment will help scientists identify the environmental factors that interact with the genes influencing educational attainment.

As we emphasized in our book *Troublesome Science*, this latter was something of an overstatement:

> We suggest that, while Dr. Harden's intentions are admirable, there are some basic misunderstandings about genetic architecture she has adopted. Dr. Harden is at pains to dispel eugenic intentions in her editorial; but alas, the specter of this 800-pound gorilla in the room can never be eliminated when the matter of the genes is raised in any discussion of intelligence. And intelligence is far too slippery a concept to be discussed usefully in the context of genes, consisting as it does of numerous different forms of acumen each of which doubtless has a very complex genetic underpinning. What is more, all those capacities are distributed across the population in noncorrelated ways.

Adding more genes to the list is clearly not a viable way out of the problem. We are thus compelled to conclude that variations in performance on tests assessing intelligence or educational attainment cannot be usefully addressed by associating yet more genes. We need to stop using reductionist approaches to traits like intelligence and educational attainment, and we need to stop trying to tie them to race. Yes, the attempt to raise general educational attainment is a wholly desirable one. But in achieving it we would be better off trying to assess the whole educational spectrum and its characteristics. Rather than particularizing, we should recall that phenotypes are the product of gene–environment interactions; and perhaps the most telling aspect of the study we have been discussing is that only a miniscule genetic component was involved in achievement. Overwhelmingly, we need to be looking at environmental influences.

Tough Luck

The brick wall we describe in this chapter exists because many traits involved in racial science are incredibly complex and difficult to work with. As an example, in 2018 we were asked to write a paper on the genomics of language, an undoubtedly complex human trait. Rather than just review the literature, we decided to see if, using current genomic information and informatic techniques, we could "discover" genes for language. Our approach involved scanning the genomes of modern humans and archaic members of the genus *Homo* (including Neanderthals and Denisovans) to see if we could identify any genes of *Homo sapiens* or of the genus *Homo* that might be specific for language. This exercise resulted in very limited matches, in only three genes: *CNTNAP2*, *GLI3*, and *WDPCP*. Of these three genes, *CNTNAP2* and *GLI3* are involved in neural development and embryogenesis and might have roles in the evolution of the neural regulatory network involved in language. *WDPCP* does not seem to have any relevance to language, or to neural development. It became obvious to us that "fishing" for genes using this approach (and really using any approach) would probably result at best in identifying genes that (while making biological sense, as did the original excitement over the *FoxP2* gene and later the *SRGAP2C* gene) would only peripherally be related to language acquisition.

As we pointed out earlier, we have more recently examined the potential biological/genetic bases for several human meta-behaviors, such as religious feelings, ethical behaviors, our tendencies to warfare and aggression, political preferences, sexual behaviors and identities, economic systems, and intelligence. Our analyses of these behaviors in *The Accidental Homo sapiens* forced us to conclude that:

> Our scamper through this miscellaneous selection of human behaviors shows pretty clearly that attempts to associate human genes with

particular complex behaviors have yet to get us very far; and it suggests that they are rather unlikely to get us much farther in the immediate future.

Matt Rockman and Kenneth Weiss have made similar arguments about the applicability of genomic approaches such as QTLs (quantitative trait loci) and GWAS to evolutionary studies of complex traits. After closely scrutinizing the approaches, they came to the rather depressing conclusion that either genomic approaches are using the wrong tools to approach key questions concerning complex traits, or researchers are simply asking the wrong questions. Rockman states that there is a "mismatch between question and method," while Weiss argues that we should strive for a "clearer goal." In the end, though, we simply might not need to fully atomize complex traits to make explanatory progress in evolutionary biology or medicine. That's because we have reached a point at which we can specifically acknowledge that the traits we are most interested in are complex, and immune to simplistic genomic dissection.

We can summarize this problem in two words initially used by Richard Lewontin to conclude a paper he wrote in 1998: "Tough luck." Tough luck, because he saw no way in which human cognitive function could be atomized in a genetic context. Like Lewontin, who uses those hard words to admonish those searching for the genetic basis of cognition, we suggest that in the end we will need to accept that many of those complex behaviors that are assumed to be racial have such incredibly complex genetic underpinnings that they will never be illuminated by simplistic racial categorizations. Accepting tough luck is, of course, tough. But it is tougher and a whole lot less satisfying when we continue to chase just-so stories. It also obscures the real value of racial science, which lies in determining the environmental contexts of behavioral or cultural/social patterns.

8 Human Adaptations

Human Variation

Our young and originally tropical species *Homo sapiens* has spread, in an amazingly short period of time, to occupy more areas of our planet than any other animal species has ever contrived to do. Human beings reside on all five continents, and in virtually every environment that those continents have to offer. No species could ever achieve a distribution like this without being highly flexible and responsive to environmental conditions; but whereas in other species adaptation to local circumstances is basically a matter of biology, and of behaviors very closely related to biology, the secret of humankind's success undoubtedly lies mostly in its unique possession of material culture, our technologies allowing us to make incredibly fast and flexible responses to varying local conditions. By the time *Homo sapiens* emerged, the genus *Homo* and its predecessors already had more than two million years of complexifying cultural evolution behind them, not to mention the shift to symbolic reasoning; and, as a result, there is no doubt that *Homo sapiens* had a flying cultural start when it left Africa to colonize the rest of the world. Nonetheless, it is breathtaking how fast that colonization happened: By 29 kyr ago – even as the world was significantly cooling in the approach to the peak of the last Ice Age – the descendants of the first tropical *Homo sapiens* had already reached the Arctic Circle.

Given the youth of our species, and the fact that all human societies have highly complex material cultures that significantly buffer them against the vagaries of the environment, it may not seem all that surprising that Charles Darwin found it difficult to attribute any of the many minor physical variations long noted among human local varieties to his mechanism of natural selection. The "epicanthic" skin folds across the eyelids of many eastern Asians, for example, do not seem to affect ocular function one way or another; they are just "there." Looking at such variations nowadays, it is pretty easy to see most of them as fairly random, as the likely products simply of drift in small ancestral populations; and, as we have seen, there is actually more variation within populations than among them, in virtually any feature you might care to name: You will occasionally encounter eye folds in Europeans, too. But nonetheless, although it is impossible to directly blame climatic and other causes for the groupings some have perceived as "racial," there is no doubt that in certain respects we can identify biological responses to local conditions – although on closer examination those responses often turn out to be as much a product of history as relevant to immediate circumstances. And neither do they suggest any utility in trying to organize the variety we do see in terms of racial groupings. Let's start with skin color, since it is so misleadingly emblematic of "race."

Skin Color

Variations in the color of human skin result from varying intensities of deposition of the pigment melanin in the outer layer of the skin, the epidermis. All humans have some melanin; those with the darkest skins have the most. Almost certainly the early *Homo sapiens* who were confined to the continent of Africa had dark skins, for melanin performs the vital function of shielding the delicate lower layers of the skin from the damaging ultraviolet (UV) part of the solar radiation spectrum. And UV hits much more strongly in the tropics than at temperate latitudes, because the sun's rays arrive vertically and therefore pass through less of Earth's attenuating atmosphere. Harmful sunburn is a response to excessive ultraviolet exposure, which also stimulates the skin to produce the extra melanin molecules that cause the protective darkening we recognize as tanning. Clearly, any tropical creature unprotected by a coat of body hair, and with an absence or minimum of unwelcomely heat-trapping clothing, absolutely requires dark skin at tropical latitudes; and there can be no doubt that the first hairless hominins in Africa (which we would probably classify, if we could identify them, as "early *Homo*") had, or rapidly acquired, dark skins. Lighter skins would only have emerged in the human lineage after the first *Homo sapiens* had left Africa; and nowadays paler skins very broadly track latitude, with melanin concentrations tending to decrease away from the equator. But why so?

Well, the countervailing advantage of less heavily pigmented skins has been debated for many decades, but by now it is quite broadly agreed that it must have something to do with the synthesis of vitamin D in the skin, a process that requires exposure to UV light. Vitamin D is a vital molecule for proper functioning of the human body, and its deficiency is expressed in any of several very debilitating conditions, including rickets, in which the bones fail to develop properly or become deformed, and osteomalacia, a thinning of the bones. Even schizophrenia has been associated with vitamin D deficiency. To a certain extent we obtain the vitamin D we need from our diets, but rarely enough to meet our requirements; and we obtain the balance through synthesizing the vitamin in the lower layers of our skin through interaction with UV light. In tropical latitudes with strong UV radiation an individual with dark skin will still synthesize enough to do the trick, but the same individual at northern latitudes – where UV is weaker and where he or she is much more likely to be wearing substantial clothing to protect against the colder weather – may well not. You can easily imagine that selection against dark skin might be pretty strong in the latter circumstance. Nonetheless, it may be instructive that, while *Homo sapiens* first arrived in Europe well over 40 kyr ago, DNA analysis of ancient skeletons (see Chapter 2) suggests that at least the bulk of Europeans remained dark-skinned (and blue-eyed) well beyond the end of the Pleistocene at about 12 kyr ago. In this case, it seems that history, in the form of major population movements, may have had more to do with apparent local adaptations than any changes on the spot may have done.

The molecular basis of skin color variation in humans provides us with an excellent example of how natural selection might work. The gene at the heart of skin color research is called *SLC24A5*. This gene is highly variable, and teasing apart what this variation means with respect to skin color and adaptation is not simple. One variant, found in almost all Africans, has a specific change that causes an alanine to occur in a key spot in the protein the gene produces. Many Europeans have threonine in this location. Whether there is alanine or threonine in that particular place results in proteins that behave differently, accounting for much of the average difference in skin color between people from the two continents. But if only things were this simple! Europeans and Africans are not the only *Homo sapiens* on the planet, and it turns out that 93 percent of (darkly pigmented) Africans share the alanine in that key spot with (much lighter-skinned) East Asians. Europeans and east Asians thus show similarly light skins, but via entirely different genetic pathways. The dark-skinned tropical ancestor of our species must have given rise to the lighter-skinned, higher-latitude Asian and European skin patterns using different genetic mechanisms that capitalized on unrelated kinds of genetic variation.

In addition, consider the following. Some populations of people are polymorphic for the alanine and threonine versions of the *SLC24A5* gene. People homozygous for the alanine type have the most melanin in their skin, while those individuals with two copies of the threonine type have less melanin in their skin. Sounds simple, once again, right? But the "simple story" trap applies here, too. It turns out that only 38 percent of the variance in skin color can be accounted for by the melanin gene in question. The remaining 62 percent of the variance must be attributed to other genes or gene complexes, or to genetic mechanisms other than simple Mendelian genetics. Rather than continue trying to pin down a genetic basis for skin color, it would be most appropriate to conclude (1) that there are genetic pathways that can produce skin with low quantities of melanin; and (2) that there are even more ways in which intermediate amounts of melanin in skin are produced.

In the end, then, it seems that many different genes are involved in determining melanin production in the many different peoples around the world, and that some of those genetic variants apparently arose almost one million years ago (so that early *Homo sapiens* with different skin color genes may well have participated in the initial exoduses from Africa). What is more, some peoples who now have pretty dark skins may have had much lighter-skinned ancestors, so we are clearly not looking at a simple progression from dark to light. The skin is obviously a dynamic system that has been heavily influenced by a complex history, but one which is also responsive enough to the environment to determine the very broad geographical pattern that our eyes tell us we see – a pattern that, it turns out, has been realized in different ways in different places. But what we can most confidently say is that, in terms of the genes that produce what we think our eyes see, it offers no justification whatsoever for any of the classical racial divisions of mankind. And not only do those races not exist, but in genomic terms skin color turns out not to exist either, at least as a unitary character.

High Altitudes

One kind of environmental stress to which it is basically impossible to accommodate culturally is that imposed by the low atmospheric oxygen levels that prevail at high altitudes. Sure, a mountaineer can strap on a couple of oxygen tanks before tackling Mount Everest, but this is a very recently available solution and it is no solution at all for the conduct of daily life on the world's highest inhabited plateau, at elevations of 11,000–16,000 feet. Humans are fueled by oxygen, and at those altitudes oxygen concentrations may be up to 40 percent lower (and UV radiation some 30 percent higher) than they are at sea level. Just for reference, Federal Aviation Administration rules make it mandatory for pilots to use supplemental oxygen when flying at altitudes of over 10,000 feet. The only way to cope with such extreme conditions on a day-to-day basis is through physiological accommodation; and it turns out that all modern human high-altitude populations show specializations for delivering adequate oxygen to their body tissues. Significantly, though, this has occurred in different ways in high-altitude populations in different parts of the world.

The highest populated region of all is the rugged Tibetan Plateau, to the north of the Himalaya mountains. Modern humans are thought to have lived there for several tens of thousands of years. Initial research on how the people of the region cope with life at an average elevation of nearly 15,000 feet focused on their physiology, and it was discovered that the main response of Tibetans to low oxygen levels is to increase both their breathing rate and their production of nitric oxide, a gas that also makes Viagra work by dilating the blood vessels and thereby increasing blood flow to the tissues. A whole-genome analysis of 27 Tibetans published in 2017 provided further information. It confirmed the close genetic relatedness of Tibetans to neighboring Chinese populations, and also implicated five separate alleles in high-altitude adaptation. The two genes showing the strongest signs of selection were *EPAS1* and *EGLN1*, both of which were already suspected of involvement in high-altitude adaptation among Tibetans; two others were specifically associated with low oxygen levels, and the fifth was a variant of a gene involved in vitamin D metabolism that is thought to assist with the deficiency of this vitamin that is chronic among Tibetan nomads.

EPAS1 is triggered when oxygen levels drop, causing an increase in the production of red cells in the blood and of the oxygen-carrying molecule hemoglobin. This can be an advantage at low altitudes when quick bursts of activity are required, such as among some athletes. But when you are living continuously at high altitudes it can be a problem because those same effects thicken the blood, potentially causing long-term circulatory problems that include hypertension and heart attacks. The Tibetan variant of *EPAS1* suppresses the rise in hemoglobin and red cells when an individual is chronically exposed to the low oxygen levels prevailing at high altitude, and thus avoids those deleterious effects. The Tibetans, it turns out, are extremely efficient users of oxygen.

But perhaps most intriguingly of all, the Tibetan form of *EPAS1* turns out also to have been present in the mysterious ancient Denisovans; and the suggestion is that this gene found its way into the Tibetan population by interbreeding prior to the extinction of the latter at some point after 40 kyr ago. Given that the genomic study concluded that the Tibetans and the Han Chinese had already been differentiating well before that time, this inference seems entirely credible. It seems, then, that the Tibetans' ability to occupy their difficult habitat may have been made possible, and was at least facilitated, by the essentially chance event of encountering a now-extinct human relative. Making this event even more plausible is the recent discovery at a site on the Tibetan Plateau of a 160-kyr-old mandible identified as Denisovan, at an altitude of more than 10,000 feet; and more recently still, Denisovan mtDNA was extracted from sediments at another Tibetan locality lying at close to 11,000 feet. Those sediments are dated to 100 kyr and 60 kyr ago; in combination, the two sites provide clear evidence that the Denisovans themselves were able to tolerate extremely high altitudes.

The high-altitude humans with the longest history of study are the inhabitants of the Peruvian Altiplano, which reaches heights of around 14,000 feet. Back in the sixteenth century, invading Spanish conquistadors complained of constant headaches and muscle weakness up on the Altiplano, and it was centuries before the first Spanish child was born there. Even today, babies born on the Altiplano to mothers of Spanish descent have lower birth weights than those born to local Quechua mothers – though both weigh less than their counterparts born at sea level. Yet the Altiplano was the nerve center of the great Inka Empire, and its native inhabitants carried on their rigorous daily business unruffled. Early studies found that the Andeans achieve this by allowing the levels of hemoglobin in their blood to rise (and thus avoiding problematic side-effects by some other means), and by developing larger lung capacities relative to their sea-level relatives.

A genomic study of the skeletons of several individuals who had lived near Lake Titicaca between about seven and two thousand years ago predictably failed to find any equivalent to the Tibetan *EPAS1* variant, but it did turn up a gene, known as *DST*, that is related to heart and cardiovascular function. The suggestion is, accordingly, that greater efficiency of the heart muscle accounts at least in part for the superior abilities of the highlanders to deliver oxygen to their body tissues at altitude. Interestingly, though, a stronger signal was found in genes associated with the digestion of starch, plausibly as a response to a diet that depended heavily on the potatoes that were initially domesticated in the Andes. And finally, the investigators picked up evidence in modern Andeans of an immune receptor that is related to smallpox resistance: a reminder that initial contact with the conquistadors was followed by a wave of smallpox epidemics, and that it is the descendants of the survivors who populate the region today. Selection has clearly been at work locally in these features that are so closely related to survival.

In Africa, Ethiopia also offers another opportunity to compare people living at high altitudes (in this case, above 8,200 feet) with those at lower ones; once again, different mechanisms seem to be at work. Researchers there reported in 2012 that "variants associated with hemoglobin variation among Tibetans or other variants at the same loci do not influence the trait in Ethiopians ... we find a different variant that is significantly associated with hemoglobin levels in Ethiopians." Where a particularly strong signal was found in this case was in genes that played "a clear role in defense from pathogens, consistent with known differences in pathogens between altitudes." None of this did a whole lot to clarify what exactly the highland Ethiopians *were* doing to nourish their body tissues in their low-oxygen environments, although hemoglobin control was implicated; but the final conclusion was clear "Ethiopian and Tibetan highlanders adapted to the same environmental stress through different variants and genetic loci."

It appears, then, that human populations are indeed biologically responsive to altitudinal stress. But they adapt via different pathways in different places. And the effects are intensely local: The changes we see are as visible between neighboring populations at different altitudes as they are across the continents. There is no way in which we can co-opt these data in the service of recognizing anything that can conveniently be called "races."

Lactase Persistence

Whenever the subject of human adaptation comes up, someone is bound to raise the issue of adult lactose tolerance. Humans are mammals, nourishing their young with milk. Milk contains the sugar lactose. After ingestion, this large sugar molecule needs to be broken down into its simpler component sugars, glucose and galactose, in the recipient's intestines. This breakdown is accomplished by a specialized enzyme called lactase-phlorizin hydrolase, or lactase for short. The inability to break down lactose is what is known as lactose intolerance, and it has some very unpleasant digestive consequences when milk is ingested, including cramps, nausea, bloating, diarrhea, and vomiting, all adding up to a very debilitating condition.

All infant mammals have evolved to produce lactase, but after weaning production of this enzyme usually ceases because the average mammal is unlikely ever to encounter milk again. This cessation is typical of modern humans too, with some 65 percent of the worldwide population being lactose intolerant as adults (see Figure 8.1 for distribution by country). But members of some human populations continue to produce lactase throughout life, and are thus known as "lactose tolerant," or "lactase persistent." Unsurprisingly, such people are strongly concentrated among cattle-raising populations, who have become dietarily dependent on dairy products. Under certain circumstances at least, they may well have had an advantage relative to the practitioners of other settled lifeways because, where it can be practiced, dairy farming tends

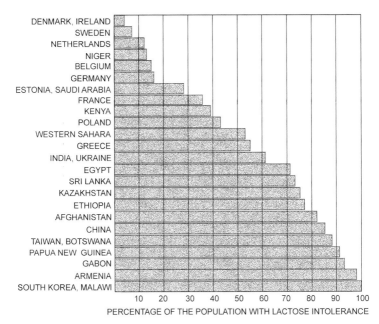

Figure 8.1   The ability to tolerate lactose (a milk sugar) in adult life is an unusual trait among mammals, but it does occur in humans. It is, however, exceedingly unevenly distributed geographically, being closely associated with populations that depend on dairy-based diets. This chart shows the percentage of native individuals expressing lactose tolerance in various countries around the world.

to provide a steadier supply of nourishment than boom-and-bust crop husbandry does. And fresh milk often provides a healthier source of hydration than potentially parasite-infested water.

In northern Europe, depending on locality, a maximum of about 15 percent of individuals are lactose intolerant, and the lowest frequencies of such individuals are found in populations that are very heavily dependent on dairy foods. Hence, the Basques come in at a minuscule 0.3 percent lactose intolerant and the Dutch at 1 percent. Among Europeans, lactase persistence appears to be the result of a single nucleotide change in a gene known as *LCT*, which is expressed in the small intestine. Several years ago researchers analyzed a group of Neolithic human skeletons from western Europe, between five and six thousand years old, and found that every individual had lacked the lactase-persistent variant of *LCT*. Ancient western Europeans were thus lactose intolerant, and subsequent selection for that variant must have been extremely strong to have achieved the high frequencies observed today, suggesting significant survival advantages. It is unlikely that those advantages would have applied in the absence of a steady milk supply, and the researchers constructed a model whereby dairying and lactase persistence started to co-evolve around 7,500 years ago somewhere between the Balkans and central Europe, and spread slowly outward from there, accounting for the single genomic variant and the quite late persistence of lactose intolerance in the west.

Outside Europe, the lowest frequencies of lactose intolerance are found in Africa, where cattle herding has a long history, albeit a spotty distribution. Cattle herding populations in Africa typically show around 20 percent lactose intolerance, whereas in populations that do not depend on cattle it can come close to 90 percent. The highest rates of lactose intolerance are found among native Americans, who are said to come in at nearly 100 percent, and among the Chinese at around 95 percent. Among East Africans, an initial study found that adult lactose tolerance is associated with three different, and independent, single nucleotide changes in the LCT gene; a follow-up study on a wider sample of Africans turned up two more. The lactose-tolerant phenotype thus evolved multiple times in Africa, and independently in Africa and Europe. Gene exchange between adjacent populations with differing lifeways has blurred the lines a bit, but it is nonetheless pretty clear that there is a fairly tight association in Africa between high frequencies of lactase persistence and diets high in dairy products. Cattle herding in Africa extends back in time for 10 kyr at most, suggesting that, as in Europe, selection for lactase-persistent phenotypes has been very strong where it occurred. This would not be unexpected in a trait that is as closely connected to individual survival as the ability to tolerate a major dietary component clearly is.

So, once again, while we find variation among populations in an important genomic and phenotypic characteristic, we do not find it distributed in such a way as to suggest any utility to "racial" classification of the populations involved.

Athletic Prowess and Body Shape

Many of the traits that appear to distinguish people from different places are hard to measure. One exception is running, where times are kept to the split second, and some anatomical and physiological measures seem to be correlated with performance. For example, compared to the rest of us, sprinters tend to have leg muscles that are rich in fast-twitch fibers that contract rapidly but fatigue relatively quickly compared to the slow-twitch fibers that burn oxygen steadily and can repetitively contract for a longer time. Some Penn State researchers asked back in 2009 whether there were any anatomical differences matching this difference in muscle architecture, and they concluded that there were. They found that, compared to a less active control group, a sample of a dozen sprinters had longer toes and shorter shin and heel bones. The longer toes were unsurprising, since they give the foot more prolonged contact with the ground as it extends, much as do the nonretractile claws of cheetahs, the fastest mammals. The short heels were harder to explain, as a longer heel would provide better leverage for the calf muscles, but it turned out that in the sprinters the contraction of those muscles was shorter, as the slow-twitch fibers were also recruited to pull on the heel. The fast-twitch fibers helped with the speed, but the added power of the slow-twitchers was necessary for optimum performance. The bottom line was that your prowess as a sprinter is indeed "in your genes," and that no matter how much you practice, you are unlikely to make gold if you don't have the requisite anatomy.

As it happens, for quite some time a preponderance of the world's fastest sprinters has been of ultimately West African origin, while eastern Africa has produced many of the top long-distance runners. We don't know if the distance folk have any typical conformations of the lower leg musculature and bones comparable to those reported for the sprinters (or, indeed, if the top sprinters themselves conform), and there is a chance that such differences might be found. But even if they were, would that mean that any such study would be more informative if its subjects were classified by race? Almost certainly not. For a start, under American rules any participants with African antecedents would be classified as "black," confounding two groups with stellar but entirely different records of athletic achievement. What's more, each group has blurry edges that are only becoming more so. And if we can learn anything from this example, it is that if you are not very careful, what you discover may well turn out to be an artifact of how effectively or realistically you define the group concerned. "Sprinters" is a pretty good category, because it is limited and straightforward: You are a sprinter or you are not (how good, of course, is another matter). But "black," or even "sub-Saharan African," would fail the utility test because of excessive vagueness, and because of its use of an *a priori* classification with no necessary relevance to the issue at hand. And, in any event, it is impossible to know to what extent transient cultural or environmental influences might be involved in determining prowess in different sports. It was different yesterday, and it may well be different tomorrow.

Concluding Thoughts about Human Adaptation

In any survey of human adaptation, it is necessary to note that some human populations living at various latitudes very broadly conform to a set of "rules" that zoologists have noticed apply to mammals in general. Allen's rule, for instance, states that within a particular mammal group, body form will tend to be linear in hot climates and more compact in cold ones. Very broadly, this applies to humans; for example, the equatorial Maasai people are typically tall, with slender trunks and limbs, while the Inuit of the New World Arctic are much shorter and more stockily built. The explanation for this difference is simple, and it relates to the physiologically critical issue of maintaining body temperature. The big environmental problem the Maasai face in their hot, dry tropical home is one of shedding unwanted body heat. Humans lose heat by the evaporation of sweat from the skin all over their bodies, and a linear build is ideal for this, maximizing the heat-losing surface area of the body relative to its heat-generating interior. In contrast, if you live in a very cold environment you will want to conserve body heat as much as possible, and the ideal form for that would be a sphere, with the smallest possible surface-area-to-volume ratio. Inuit are, of course, no more spherical than the Maasai are cylindrical, but their stocky build clearly helps with that heat conservation. Appropriate cultural accommodations to their respective environments are also made by both groups, but there is no doubt that in both cases their bodily adaptations help a lot.

But, yet again, although we clearly see adaptation at work in such fairly rare cases, what we do not see is anything that would be helpfully illuminated by introducing ideas of race. The Maasai are surrounded by populations shorter and chunkier than they are, and indeed, a bit farther to the west, at the very same latitude, you will find the Central African forest peoples, among the shortest on the planet (and with good reason: in the forest being tall is a big handicap). Once more, we are looking at human variety at a level of complexity that makes a mockery of those racial divisions.

And finally, while we have seen in this chapter that it is indeed possible to find within the human species variations that do seem to be related to adaptation, it is necessary for the sake of balance to emphasize two other things. First, that the vast majority of the variations we perceive, both among individual human beings and between human populations widely separated on the planet, are best explained as random products of biological processes. Geographically their distributions, if any are clearly discernible, are often governed by the myriad population movements documented in Chapter 2. And second, as the evolutionary geneticist Richard Lewontin so eloquently pointed out, the amount of variation that exists within groups vastly exceeds that which exists between them, in whatever characteristic you might care to think of. What's more, those variations are completely uncorrelated. Your electrician's skin color, or stature, or accent will not help you in the least to predict how good a job he or she will do. And which abilities you will value most will be entirely situational: There is no way abstractly to rank mathematical ability, artistic talent, oratorical skill, or street smarts, once again things that are entirely unpredicted by any aspect of external appearance. Indeed, such cognitive traits are the very last place in which to look for adaptation.

Our brains are multipurpose machines that are constantly remaking themselves, and that are exquisitely responsive to circumstance. Contrary to what evolutionary psychologists suggest, people have not been molded by evolution to any particular purpose. Yes, some humans are "smarter" than others, or better at doing certain things. That is only to be expected. But although "intelligence" is a prized quality that has repeatedly been mentioned in connection with "race," it is something that has proven impossible to measure satisfactorily, not only because it is so complex and multidimensional, but because it is so tied up with culture, social stratification, and economics. Notions of race do nothing to help clarify this complex situation.

9 Science, Pseudoscience, and Race

"Race Science" in Context

Research prior to human genomics opened several doors to race-focused research, but the availability of genomes was a game-changer. As we pointed out in Chapter 4, there is nothing inherently wrong in using racial or ethnic boundaries as research tools for trying to discover biological patterns within the human species – if they are justified. But this is a powerful caveat, because the racial/ethnic boundaries that researchers use must be meaningful and biologically *real* if they are to lead to meaningful conclusions about our species; and in this chapter we examine whether or not that is the case.

We are not, it must be said, arguing for the complete abandonment of stratification procedures in science, for we can see some cases in which stratification by social group or ethnicity in research has been quite useful. For instance, the recent COVID-19 pandemic, while horrible in all walks of society, impacted people of color more severely than others. We know this because researchers stratified demographic data into ethnic categories, and the correlation was clear. But this doesn't mean that there are genomically or genetically defined races. We hope to show that modern racial stratification is a poor tool for understanding anything but the environmental components of disease and phenotypic traits.

In 2012, the science journalist Nicholas Wade published a book, *A Troublesome Inheritance*, that purported to use genetics and genomics to define races within *Homo sapiens*. We feel it useful to critique Wade's book here because, while we find Wade's arguments scientifically unsound, he did a very thorough job of articulating the hypotheses and approaches of the "racial science" he espouses, and his exposition clearly reveals seven major scientific errors:

- a misunderstanding of the scientific process in general, and of hypothesis testing specifically;
- a lack of appreciation for the precision of taxonomy, and of the meaning of taxonomic terms;
- a misunderstanding of what cluster analysis does;
- a failure to recognize bias and cherry-picking in data-reduction techniques;
- being fooled by the "underlying correlation structure" of human traits and genes;
- conflating geographically skewed allele frequency differences with ancestry; and
- conflating geographically skewed allele frequency differences with adaptation.

Most of these mistakes have been around for a long time in racial science, some for a century and a half. But what Wade did was to apply them to genomic data and genome analysis in the context of racial science. And he started a trend, because more recent treatments of genomic data in the name of racial science have committed the very same "sins." Thus, Charles Murray's recent books, published in 2020 and 2021, *Human Diversity: The Biology of Gender, Race, and Class* and *Facing Reality: Two Truths about Race in America*, respectively, discuss the role of genetics in establishing race as an introduction to a diatribe on racial differences in cognitive ability (a topic we discussed in Chapter 7). John Fuerst's "book," *The Nature of Race: The Genealogy of the Concept and the Biological Construct's Contemporaneous Utility*, takes a more expansive look at the history of the term "race," and offers a skewed set of conclusions in support of the proposition that modern genomic data "prove" the existence of races. Skewed in which way? One only need note the websites from which this pseudo-book can be downloaded – Metawiki, Thuletide, and Notpoliticallycorrect, among others – all of which are far-right and white-nationalist in nature.

Epistemological Pain: Misunderstandings of Science

Before we get into detail about the mistakes that are typically made when genomics is applied to race, we feel compelled to look at some of the philosophical objections that have been raised to racializing science or to racial realism. In 2015, Quayshawn Spencer examined the rush to "racial realism" from a philosophical standpoint, as a result of what were then several human genomics studies that, purportedly, visibly showed racial grouping. To clarify the argument, Spencer pointed to the following problems with rejection of, or objection to, the racial results. The first two of these problems are semantic, and the latter two are what he calls metaphysical:

- The "discreteness objection" suggests that no level of human population structure can show racial groupings, because the hypothesis that a race exists means that we need to test for discreteness. This objection is semantically faulty because discreteness cannot be a consideration in a species in which many people identify as more than one race.

- The "visibility objection" maintains that no level of human population structure can show racial groupings because those groups should be visible in distinctive traits. In philosopher's Joshua Glasgow's words, it is "conceptually non-negotiable, that each race by and large has a distinctive set of visible traits." In other words, visible traits and potential racial groups do not overlap. This argument is semantically faulty because members of the same race do not necessarily look alike.

- The "very important objection" maintains that no level of human population structure contains real racial groups because to be biologically "real" such groups need to be important enough to be classified biologically. As we have seen, though, the taxonomic term "race" is sufficiently subjective for such groupings to be rejected. Richard Lewontin first used this argument to reject biological races, and we have similarly used it in Chapter 6.

- The "objectively real objection," which states that no level of human population structure is biologically real because those levels are not objectively real. Why? Because they would not exist if it were not for human interest, belief, or some other human psychological state. Spencer rejects this objection as follows: "in order to accurately capture the real entities of human biology, we need to embrace biologically real entities whose existence is dependent on human mental states, such as human beliefs and human interests." In other words, we need the products of human mental states to formulate hypotheses.

We agree with Spencer that the rejection of racial realism cannot be based on faulty logic. Spencer's refutation of these objections by the opponents of "racial realism" are interesting, but we feel that there is conflict here with the scientific process of hypothesis testing that we discuss next. In addition, as we argue below, faulty conclusions result when there is a misunderstanding of the practice and goals of taxonomy. And finally, Spencer's arguments assume that the data actually show discrete differentiation, which is not necessarily true.

We also feel compelled to point out one other problematic aspect of publications such as Murray's *Facing Reality: Two Truths about Race in America*. Note that the word "reality" in Murray's title reflects an absolute claim for the contents of his book. Almost a decade earlier Wade made the same claim, but in a differently patronizing (to scientists) way. His argument is that, if something is scientifically sound, then we are wrong to ignore it. We could not agree more with Wade on this principle, but sadly his exposition of the genomic data as demonstration of the existence of races is seriously flawed. Still, like Wade, many who have jumped into the fray have done so as self-identified "race realists," using an elaboration of his argument that goes like this: "You scientists know that race is real from the data, and your tendency to ignore it is unscientific." But while it is true that ignoring accurate scientific results and inferences made from them is to willfully abandon scientific process, those results need to be accurate in the first place, which is where all the race realism arguments fall flat. The conclusion that races are biologically real is simply false, and to continue to test this unequivocally false hypothesis is problematic.

The first and foremost mistake that race realists make, then, is to claim that they are doing or invoking science. However, as described earlier, science proceeds by hypothesis testing. And while race realists will argue that they have indeed tested a hypothesis – "H_0 = races exist" – this is actually not the case. To demonstrate this, let's look at the methods used by researchers dealing with racial genetics and genomics and see how the "races exist" hypothesis is tested (or not).

Mistake 1: Misunderstanding hypothesis testing. We are ultimately looking for a scientific treatment of genetics, genomics, and race. One way in which science proceeds is by hypothesis testing. A hypothesis is posed, tested in the most severe way possible, and an inference is made based on the results of the test. If a hypothesis is rejected, the next step is to come up with an explanation for the rejection that the researcher can then use to reformulate research questions so that progress can be made. None of this means that one cannot revisit a previously rejected hypothesis, and in fact the revisiting of hypotheses is a healthy aspect of true science. To many this procedure might seem simple, but the practice is rigorous, and sometimes grueling.

The hypothesis that we are faced with here is a simple one – H_0 = *On the basis of biology and genomics specifically, there are X number of races of humans on our planet*. This is a racial hypothesis, but not a "racist" or a "race reality" one. It is perfectly valid, and perhaps even interesting. So, we have our hypothesis, and our next steps are to move forward and rigorously test it. But before getting there, we need some context. This is usually provided by what is called "background knowledge," the information that is relevant to posing and testing the hypothesis. Without it there is no way to anchor a scientific test. So, what is the needed background knowledge for the hypothesis at hand?

First, we need to define the term "race," and determine how we would detect it using a hypothesis test. We need a strong definition before we can go any further, one that is both objective and operational, meaning that we cannot use a biased definition or one that we cannot test. Fortunately, we know what a species is, giving us an objective and for the most part operational definition that allows us to test hypotheses about species delimitation. In fact, species recognition is an important and vibrant field of evolutionary biology, and while it has given us a couple of dozen definitions of species for sexually reproducing organisms, they are all pretty much based on the same thing – reproductive isolation. Two entities are separate species when reproduction ceases between them. But here's the rub: While you can test a hypothesis of species existence, there is no corresponding test for the reality of subspecies.

There simply can't be; a subspecies can never be more than a hypothesis, because there are no objective criteria for calling something a subspecies.

Mistake 2: Misunderstanding the biological meaning of taxonomic terms. Testing a hypothesis of species existence with genetics is pretty straightforward. You collect organisms, propose a hypothesis of species existence among the samples you collected (e.g., the ones east of this river are one species and the ones on the west side of the river are a second species), obtain more data, and test the hypothesis. The simplest test is one for diagnosis. This test simply looks for traits in the organisms being examined that are fixed and distinctive in one of the hypothesized species, and not in the other. This test can be relaxed in several reasonable ways so that it can be done under different assumptions or definitions as to what a species is.

The next taxonomic level down is the subspecies. This epithet was created to "flag" groups of organisms in nature that might be species, but that haven't been tested yet. In other words, subspecies are hypotheses of species existence that await testing. Often they are not tested, because researchers are satisfied with the subspecies epithet which, while not entirely precise, may still allow a researcher to carry on work on the group. That's because not all researchers are after classification and taxonomy, but rather simply need enough precision to formulate other hypotheses about the ecology, behavior, evolution and so forth of the organisms of interest. The term "race," which was created historically to accommodate systems in which the entities are locally adapted to their environments with unique traits but are not diagnosable, is generally equivalent to the subspecies epithet. And because it is so close to the definition of subspecies, many researchers forego its use. While they are familiar with the term, they rarely, if ever, use it in their research.

In contrast, race is extensively used in the context of humans. And while we find such usage useful in some ways, especially in social science-oriented endeavors, it is clearly very unsatisfying in others, principally because there is no objective definition of what a race is. Unlike species, in which the definition is objective and operational, the definition of race is not. It can mean many things. Unlike the subspecies concept, which is best used as a hypothesis for testing of species boundaries, race in humans is rarely used as a hypothesis.

Another term relevant here is "population." A population is a researcher-defined outbreeding entity. It can be defined geographically, anatomically, behaviorally, or otherwise, but however defined it can breed with other populations within its species. To us, this sounds a lot like a race (which is the only utility we can think of for that term). Many writers have suggested that biologists who do not like the term "race" have simply shunned that term and substituted "population" for it. We do not accept this synonymy, but both terms have in common that they are unsatisfying. This is because neither has objective criteria to define them, as a species does. One can take individuals from a well-defined species and use various methods to determine the degree of isolation certain individuals have from each other, and in this way cluster them into populations; but, again, the criteria for the clustering are subjective.

As we say above, to us race sounds a lot like both "subspecies" (which are best thought of as hypotheses) and "populations" (with their researcher-based definition of cohesion), and hence it is caught between a rock and a hard place. We see the difference between population and subspecies, but we find it difficult to objectively discriminate between subspecies and race, and population and race. This problem is probably what makes race such a misused and misapplied term in science.

So how did this imprecise genetic term make it past the discerning eyes of the field? Our colleague Michael Yudell has a notion of why. In his book *Race Unmasked* he points toward the role of the eminent evolutionary biologist Theodosius Dobzhansky. Yudell argues that Dobzhansky's 1937 text *Genetics and the Origin of Species* had a major role in getting biologists to abandon the concept that races were real, typological, entities (i.e., that they were diagnosable) and to embrace a less objective understanding of the term. Yudell argues quite forcibly that Dobzhansky's advocacy produced an intellectual struggle among evolutionary biologists, systematists, and taxonomists to define race genetically, and not be racist in doing so. As Terrence Keel says of Yudell's thesis, the scientists who followed Dobzhansky fell into a "contradictory space" (that rock and a hard place) as they "struggled to both find meaning for a race concept in science and fight against racial science and racism more generally." As an example, while Dobzhansky led the effort to be more biologically precise about race, he was an avowed anti-racist. Keel also points out that Charles Darwin was trapped in nearly the same tight spot. While he used a race concept in his writing, he was an avowed abolitionist. This paradox is at the heart of any misunderstanding about the role of the term race in biological and evolutionary studies. This paradox also complicates any discussion of whether races are scientifically real. If you follow our discussion here, then looking for races in humans is illogical – for if we don't have a viable, objective definition of race, then how can we test for it? End of story; racial biology becomes inscrutable and useless. But let us continue with some of the mistakes made by race realists like Murray, Wade, and Fuerst.

Mistake 3: Misunderstanding the meaning of cluster analysis. As we showed in Chapter 6, not only is cluster analysis used by human genomicists to present results, it is also used to support racial interpretations of those data. We pointed to John Novembre and Benjamin Peter's statement about the visual appeal of these approaches, and it is worthwhile to repeat their caveat here concerning use of clustering tools and human population structure: "A final precaution, and one of broader societal relevance, is that a viewer can become misled about the depth of population structure when casually inspecting visualizations using methods such as PCA, ADMIXTURE, or fineSTRUCTURE. For example, untrained eyes may overinterpret population clusters in a PCA plot as a signature of deep, absolute levels of differentiation with relevance for phenotypic differentiation." While these analytical methods are widely used by researchers who study human variation, they do not, as we hope we have shown in Chapter 6, test hypotheses about the biological existence of human races. Cluster analysis cannot be used to confirm or prove the existence of human races. A valid test of such grouping of individuals (which is inherent in the detection of races based on genomic data) would be phylogenetic tree analysis, because of the strong objective arbiter of monophyly or restricted common ancestry; and in Chapter 6 we described the results of phylogenetic analysis of sequences from the 1000 Genomes Project, finding that monophyly of groups relevant to a racial inter-pretation of the data are entirely absent. Indeed, regardless of whether the data are analyzed separately or concatenated into a single data set, monophyly of pre-described racial groups is lacking.

Mistake 4: Lack of consideration of bias when using data-reduction tech-niques, or when cherry-picking data sets. It should be clear that genomic data sets can be ascertained in different ways. One problem with ascertainment is that bias can be introduced into the data set, depending on the way the data are ascertained. At one end are ancestry informative markers (AIMs) and at the opposite end are those methods that consider the entire genomes of the humans in a study. In other words, the spectrum ranges from extreme ascertainment bias (AIMs) to no bias (full genomic consideration).

Ancestry informative markers and even some of the more popular human variation panels for analyzing human variation are ascertained with what we call a "flawed white swan approach." To explain, we first need to know that Karl Popper, the philosopher of science who was so forceful in articulating the hypothetico-deductive approach to scientific inference, used white swans to demonstrate problems with inductive thinking. The statement or hypothesis "All swans are white" can be falsified. How? By simply finding a non-white swan (there are indeed black swans living in different parts of the world). Popper would call this statement a reasonable hypothesis simply because it has the potential for being falsified. AIMs, on the other hand, are not white swans, because they are ascertained by surveying a subset of the variation in populations, taking those variant sites that are diagnostic or at high frequency for a group of interest, and then using these to define a group. This approach is equivalent to hypothesizing that "all white swans are white." This approach to Popper is not scientific, but inductive – an approach in science that Popper warned against.

The result of this way of looking at tests of the existence of human races and at human genomic variation reminds us of the comedian Stephen Colbert's response to the results of his "ancestry" test on the PBS show *Faces of America*. Subjected to a version of the AIM approach, Colbert was informed by a pie diagram that he was 100 percent Caucasian, upon learning which he facetiously characterized himself as the "black hole of white people." This was absolutely not the case, however, because the majority of his genome is African in origin. A common argument used against this criticism is known as the "Mount Everest Paradox." The argument goes as follows: The elevation of Mount Everest differs from the surface of the ocean by an incredibly small fraction (about 0.0008) of the Earth's diameter. But anyone standing at the foot of Mount Everest can tell the difference, and it is huge. Again, this is a trivial and unscientific argument: One could just as easily argue that, to a bacterium, a golf ball looks like Mount Everest (indeed, a 0.0002 percentage diameter-wise). Well, yes, but any golfer can tell you how hard it is to find a golf ball in the rough, because it is so small. It is not the changes or differences that matter; but rather what the differences mean, and whether or not there is an objective way to interpret them.

Mistake 5: Being fooled by the "underlying correlation structure" of human traits and genetics. Many researchers fully recognize the mistakes we have discussed so far. Unfortunately, even when researchers do recognize the nuances of hypothesis testing and the problems with tree building and cluster analysis, they continue to use them anyway. Why is this? The statistician J. C. Gower had a good answer: "the human mind distinguishes between different groups because there are correlated characters within the postulated groups." The underlying correlation of data that Gower mentions brings us to how researchers have used treeing approaches and clustering approaches to address biological race, and it is those correlated characters to which we now turn. Gower's "correlated characters of the postulated groups" were what A. W. F. Edwards used to argue against Lewontin's nonetheless correct observation about human variation. In the 1970s, Lewontin examined the available data for human variation in blood groups and some proteins. He concluded that there is more variation within a postulated racial group (85 percent of the total) than between such groups (15 percent of the total). Lewontin's observation was a strong argument for the abandonment of genetic racial divisions.

In a similar approach, Lynne Jorde and colleagues asked the question "How often is a pair of individuals from one population genetically more dissimilar than two individuals chosen from two different populations?" This question is just another more operational way of assessing Lewontin's claim. As we implied in Chapter 6, the answer depends on the number of loci (or, in genomics, the number of single nucleotide polymorphisms, or SNPs), and can be as much as 30 percent for low numbers of loci and limited population sampling. Jorde and colleagues point out that rare allele loci will result in a high frequency of between-population similarity, greater than the within-population frequency, and that with ample population samples and large numbers of loci the answer to the question will never be zero. This result suggests that Lewontin was more than likely correct with his original assessment in the 1970s.

What, then, is Edward's issue with Lewontin's claim? It is quite simple, really; he claims that Lewontin's original 1970s conclusion ignores the correlation structure of data, and in so doing misses the fact that populations can be distinguished and delineated from each other using that underlying correlation structure. If this sounds like data reduction, then it should. We humans have the tendency to find ways to group things based on our notions of what *we think* is important. As residents of New York City, we both ride the subway to work frequently (until the pandemic that is). Ridership on NYC mass transit is quite diverse with respect to the outward phenotypes of people. This diversity can be mind-boggling at times, and the human brain craves some way to make sense of it. It is quite easy to fall into the trap of putting people in a subway car into groups, based on some underlying notion of group membership. The underlying notion is the hidden correlation structure we have been discussing. But we hope you can see the futility of this endeavor. Two unrelated people with a particular skin color (which would no doubt be the first phenotype we would use) cannot necessarily be put in the same group based on genetics or sociocultural origin. Skin color, while having a definable range of genomic contexts, does not translate into a systematic phenotype for the classification or grouping of people.

To claim that a diagnostic exists for a group of people from the underlying correlation structure of that group of people is unscientific in many other ways as well. In essence, due to the amount of genetic variation the human genome contains, we could probably take a roomful of students from a typical university class, sequence their genomes, and find SNPs that would trick us into thinking that they are distinct from other students on the campus, or find a SNP or two possibly diagnosing those students as distinct from other humans on the planet. We could even do this for related members of a family group (this principle is what paternal DNA tests are based on). As a result, we see the underlying genetic genomic correlation structure as an extremely subjective data-reduction methodology that only presents us with more hypotheses. Those hypotheses then need to be tested with more rigor than was used to pose the hypothesis in the first place.

Since some of the issue of biological race is a systematic/taxonomic one, we feel compelled to examine what a systematist might say about underlying correlation structure. And modern phylogenetic systematic theory would not necessarily agree that a fallacy exists. But what Edwards is asking us to do is to reduce or "throw out" some of the information in the overall data set – indeed, a great deal of it. Modern phylogenetic systematics, on the other hand, would search for a signal in the data set: not only in the information that exhibits underlying correlation structure, but also in the rest of the data. And remember that the visual neatness of STRUCTURE and PCA approaches becomes more and more hazy as the number of individuals in the analysis increases.

When we examine the ontological constraints of systematic analysis, we can only conclude that underlying correlation structure is not valid for extracting hierarchy or discreteness from a data set in a systematic context. So, what good is it? We acknowledge that correlation structure exists for genetic/genomic information. We also acknowledge that it should be useful. We suggest that it is telling us something about ancestry, but that it is completely irrelevant to delineating the existence of populations within our species, or with establishing hierarchy of individuals in our species. Ancestry really should be the focus of studies using genome information of *H. sapiens* on a large scale. But an exploration of ancestry needs neither to use phylogenetic trees or clustering approaches, nor to imply grouping or hierarchy. In fact, trees and clustering approaches are illogical when the goal is to understand the ancestry of individual *Homo sapiens*. Focusing on individual ancestry, and moving away from hierarchy and clustering, also places human genome analysis on sounder scientific footing. Unscientific assumptions of group membership will be avoided, and the science will become sounder.

Mistake 6: Conflating racially based genetic differences with explanation of ancestry. People are very interested in their ancestry and lineage. The broader availability of direct-to-consumer (DTC) genetic ancestry testing has purportedly made understanding one's ancestry possible. But what do ancestry tests tell us? Simply put, they give us details about chunks of DNA in our genomes, and where they might have come from. In this context, as some authors have claimed, ancestry testing has become a proxy for race determination. This is an unfortunate development in the use of genetics and genomics, mostly because our genomes are mosaics of ancestry, even including bits of DNA that show ancestry with other species. The bottom line is that an ancestry test is more about the chunks of your genome than it is about you, or about your inclusion in any predetermined or concocted group. This chunk of your genome may have markers of Spanish ancestry, but that doesn't mean the entirety of you is Spanish. To be Spanish means so many other things, most of them nonbiological.

When it comes to our understanding of race (or the lack thereof) in humans, ancestry approaches are also flawed due to the lack of objective knowledge of how many of the variants (and even more complicated, *which* variants) make groups of people significantly different. No self-respecting scientist would even come close to designating objective criteria, because such criteria simply don't exist. As we indicated earlier in this chapter, ancestry can be traced all the way to the level of related family. Does that mean we need to start considering family lineages as hypotheses of species? Hardly. Ancestry simply connects us with one another, and it is not hierarchical with respect to any purported higher level. So what, in the end, do genetic ancestry tests tell us? A good way to view DTC ancestry testing is as "recreational genomics," and such recreational approaches are difficult to meld with our current vision of science. Some studies have used immense data sets from companies like Ancestry.com and 23andMe.com. Those studies are limited by the metadata that the DTC companies have collected along with the cheek swab or the saliva from their consumers. These studies are often retrofitted to these metadata, making the science of establishing the DTC databases hypothesis-free. While hypothesis-free science leads to lots and lots of data, those data are difficult to retrofit to hypotheses, making such studies of questionable relevance. It is even arguable whether DTC tests even offer anything (except occasional grief) to those engaged in the recreational activity.

It is often argued that some conception of race is needed if we are to study the historical movements of humans and their evolutionary history. But this is entirely false. As discussed in Chapter 5, we already have an excellent grasp of how humans migrated in the past based on mtDNA and Y chromosomes and the fossil record, without recourse to racial designations. We are not impeded at all in these endeavors by the lack of formally defined species, subspecies, or biological races. This is because clonal markers that follow individual haplotypes are used, and hence no *a priori* definition of race is applied to interpret the results in these tree-based analyses. Race has also been argued to be an important component in medicine, and the jump is then made that race is essential to the health of people. Once that jump is made, it is easy to claim that a scientist who does not consider race is impeding medical progress, something nobody would want to be accused of. Happily, medicine will soon benefit from individualized genomics, and – because race and ancestry have been poor tools in medicine – we suggest that there will soon be no reason whatever to consider race in medicine. Ancestry will likely be important in this context, but trying to insert concepts of race into medicine is truly barking up the wrong tree.

Mistake 7: Conflating geographically skewed allele frequency differences with adaptation. Adaptation and allele frequencies are the focus of Wade's last five chapters in *A Troublesome Inheritance*, and form the basis of Murray's diatribe in *Facing Reality: Two Truths about Race in America*. The apparent justification for this is that we need to have a notion of races so that we can explain why some of us look and behave differently from others. Yet nearly all the (remarkably few) "adaptations" that can be identified appear to be intensely local in their occurrence – for example, the diverse responses to high-altitude living and to living under intense solar radiation – and are not at all usefully illuminated by any concept of major "races" (see Chapter 8).

In a review of Wade's book, Charles Murray placed a bet that attacks on the book would be made on sociological lines, based on scientists' fear of breaking away from tyrannical orthodoxy. We think Murray lost his bet, as we can soundly refute the genetic/genomic arguments for race, as shown in the first five mistakes discussed above. Those mistakes aside, we point to a more insidious argument that Wade employs – the tyrannical orthodoxy. He claims to fear that unorthodox thinking tends to get stifled by orthodoxy, so that progress, both scientific and social, is impeded. Murray is, if anything, even more steamed about this. Both, however, are entirely on the wrong track. What they do not recognize is that the best science occurs when the science itself is examined and placed under real scrutiny. The raising of "unorthodoxy" to a place of eminence in the scientific process by both Wade and Murray exhibits a sad misunderstanding of how science is done. We might call Wade's insistence that science advances by departure from orthodoxy the Indiana Jones Fallacy. It is especially important to understand the fallacious nature of this idea because all the positive reviews of Wade's book (Murray's included) harped on the far-reaching importance of Wade's departure from the tyranny of scientific orthodoxy. But while it is true that the eminent philosopher Thomas Kuhn argued in his *The Structure of Scientific Revolution* that unorthodoxy played an important role in the advancement of science, he never claimed that normal orthodox science was irrelevant to such advancement. And more importantly, he never argued that unorthodoxy was equivalent to scientific revolution.

As scientists, we recognize how gratifying it might be if every published scientific paper was earth-shaking and unorthodox. If so, scientific progress would be rapid and unlimited, although virtually impossible to keep up with. But the sad truth is that much of science is rather boring and procedural – exactly as rigor demands. Even the hypothesis that there are genetic differences among people from different geographic regions – classifiable or not – is pretty mundane, since of course there are differences among humans, as there are in any widespread species. We don't need to spend millions of dollars sequencing genomes to know this. The real questions are whether the differences really are significant and/or interpretable in a rigorous scientific context, and whether the classification of people into races helps us to understand them better. In the first case, while there may be minor differences, they do not seem to sort out on larger scales. And in the second case, the answer is a resounding "No!"

The Bottom Line

It is appalling how many white-supremacist, white-nationalist, far-right, and straight-out racist groups have co-opted scientific work into the racial realm. The websites of these groups are often no more than a click or two away from searches using terms like "Darwin", "STRUCTURE analysis," or "PCA." Many of the websites representing these groups have pirated Darwin and his allusion to race in our species. In particular, we might cite the famous quote from his *The Descent of Man*: "Some of these, such as the Negro and European, are so distinct that, if specimens had been brought to a naturalist without any further information, they would undoubtedly have been considered by him as good and true species." It is true that Darwin suggested that races might be the best way to describe the status of different people on the planet, as understood in the second half of the nineteenth century. But it is obvious from his writing that he did so begrudgingly, and there are two problems with citing this quote as proof of race. First, the relevant part of Darwin's book goes on for several pages presenting that "further information" that is mentioned in the quote, and he concludes not only that all humans on the planet have a common ancestor, but that the lines between races he mentions are blurry at best. Second – and we are sure that Darwin, as a supremely competent scientist, would have recognized this – his word would not have been the last one on whether races exist. To co-opt Darwin in the attempt to prove that races exist is akin to relying solely on Antonie van Leeuwenhoek and Robert Hooke for a modern defin- ition of microbes. They had huge insights into what microbes are, but their ideas are only a small part of what we now understand about microbes. It is not a stretch to suggest that Darwin understood that race and subspecies (two terms he used in discussing our species) were hypotheses for species exist- ence. And we would hope he would recognize that the racial hypotheses we have discussed in this book are all soundly rejected.

We conclude with a quote from our own book *Troublesome Science* that we hope is not too self-serving. We just couldn't think of a better or shorter way to say it:

> For almost all of the past 50,000 years or so since *Homo sapiens* has been widely present throughout the Old World, our hunting-gathering pre- cursors were sparsely spread out across vast landscapes, and constantly buffeted by rapidly changing climatic and environmental conditions. This provided optimal circumstances for the incorporation of minor genetic novelties into local populations, and it explains why, for example, Africans generally tend to resemble each other more closely than they do Eastern Asians or Europeans. But all of us remained members of one single, interbreeding species, and we guarantee that the edges between populations were never sharp. What is more, over the past ten thousand years since the adoption of a more settled way of life, demographic circumstances have changed entirely as populations have mingled on a large scale and often over vast distances. This, above all, is why it is hopeless to look for the boundaries that are necessary if we are to usefully recognize "races." The central tendencies may be there, but the bound- aries aren't. Which means that "race" is a totally inadequate way of characterizing, or even of helping us to understand, the glorious variety that is humankind.

Summary of Common Misunderstandings

Here we list and respond to some of the common misconceptions about race that are discussed in this book.

Human races exist. The human mind has a deep desire to catalog and classify the elements of the world around it, and this also goes for the varieties of humankind that we perceive. Yet in doing so we actually distort our perceptions of the world by ignoring its awkward complexities. Division of the peoples of the world into four, or five, or even a couple of dozen definable groups has been repeatedly attempted and has proven an impossible task. Our species *Homo sapiens* is undoubtedly variable in its physical features; but it is variable in so many different ways that no system depending on defined groups can capture that variability. During the Ice Ages, small and isolated local populations of *Homo sapiens* undoubtedly acquired some minor distinctive features; but since those times the human population has mushroomed, and people have streamed hither and yon across the face of the planet, interbreeding until the species is one glorious mishmash. Boundaries, to the extent that they ever existed, have become blurred, and categories within the species impossible to define.

Different (racial or ethnic) groups have different abilities. It has often been assumed that physical differences among people must map onto different spectra of ability, or, even more destructively, of superiority and inferiority. This has proven not to be the case. Within particular societies, geographical origin may generally correlate with social or economic disadvantage, and thus with educational or economic achievement; but when such nonbiological influences are factored out, the correlation disappears. What is important here is historical factors, not biological ones. By looking at someone's biological features, it is impossible to predict with any certainty virtually anything about them.

Skin color is a useful feature for defining racial groups. Skin color is hugely variable among human beings, from the very pale to the very dark. And it is one of the few variations that has a clear adaptive basis, dark pigmentation being hugely advantageous in the tropics (where our species evolved) and mildly disadvantageous at high latitudes. Yet skin tone doesn't reliably define any larger groupings within the human species, and it turns out that both darker skin tones, and lighter ones, can be arrived at by multiple genomic pathways. There is more than one way in which to arrive at a particular skin color, and people with apparently the same skin color may have entirely disparate histories. This misconception also ignores the fact that what the observer is seeing with respect to skin color is the underlying correlation structure of the grouping done in the mind of the observer. The underlying correlation structure is what we use to visually and quickly categorize things (including people). It is not empirical, nor is it hypothesis testing. Even when we use techniques that purport to extract the underlying correlation structure (like AIMs) of a population of people, we still ignore other data, which is in itself unscientific. Some have argued that the underlying correlation structure of the genome is enough to rely upon to prove genetic races exist. This underlying correlation structure is not what a taxonomist would use to test a hypothesis of the existence of race.

The geographical varieties of humankind have deep roots in time. We have a keen eye for our own species, and so we tend to overestimate the significance of the differences in appearance that we see among people from different areas of the world. In turn, we often conclude that those differences must have been established over long periods of time. In reality, however, all biological differences among modern human beings are of extremely recent origin. Our species *Homo sapiens* evolved a mere 200,000 years ago, a blip in evolutionary time. And not only are all variations among humans around the world an epiphenomenon of that short period, but all variations we see outside the continent of Africa are a product of no more than the last 70,000 years or so. What we are looking at in *Homo sapiens* is within-species variation, and variation within a very young species at that.

Physical differences among people from different areas of the world must mean something biologically. We are often taught that evolution is about adaptation, and that all physical features must have key adaptive significance. But this distorts the process. New genetic variants arise at random; and, even if they are of no particular advantage, they may simply linger around if they don't cause problems, sometimes even becoming "fixed" in local groups. This seems to have been the case with the vast majority of the differences we see among people from different parts of the world. The adaptationist Charles Darwin was puzzled by the fact that he was unable to impute biological significance to almost any of the human variations he cataloged, and evolutionary scientists today recognize the random nature of those variations, which do not significantly affect overall function one way or another. This does not mean that a few variations were not of adaptive importance at the times and places they arose, and such variations indubitably include such highly visible variations such as skin color. Importantly, however, those variations are complex, and they are not usefully explained by any concept of "race."

We can tell if there are genetic races by just looking at patterns of variation of humans. This misconception disregards the time-tested approach of hypothesis testing in science. The hypothesis that is most relevant to examining the genetic reality of race is:

H = there are n number of human races.

where n is a number representing continents, types of skin color, etc. This is the bald-faced racial hypothesis when using any kind of data to say races exist. But how this hypothesis is tested and what the tests mean are almost always misconstrued. When one looks at this hypothesis in a testing framework it is always rejected, no matter what value one uses for n.

Arguments against the existence of genetic races are semantic. This misconception neglects the importance of taxonomic thinking and trivializes the science of taxonomy. In fact, taxonomists test hypotheses (like H in the previous misconception) on a regular basis. The designation of a group of organisms as a species, subspecies, race, or other is a taxonomic act and is precise. The rules of taxonomy should be followed in any designation of a species, subspecies, or race epithet. The science of taxonomy does not result in semantic arguments but sound scientific conclusions and precise nomenclature.

The modern techniques that are used to analyze the genomic basis of race, such as STRUCTURE and PCA, clearly show the genetic existence of races. This statement is clearly a misuse of these statistical techniques. The misconception conflates the results from genetic or genomic analysis using clustering methods like PCA, STRUCTURE, or tree building as being relevant to testing the hypothesis that races exist in a genetic context. The most precise method for testing the cohesion of a group of organisms, or the lack of cohesion, is through tree analysis. When this approach is used it fails to tell us anything about race. More importantly, clustering approaches like STRUCTURE and PCA are best used to visually summarize data and pose new hypotheses, and should not be used to test hypotheses.

Ancestry means race in humans. Because all humans are an interbreeding group of organisms, hierarchy of the relationships of individuals is not present and hence any grouping of people is not possible. In reticulating or interbreeding groups of organisms, the genealogical hierarchy is lost as genomes recombine through sexual reproduction. Instead, what relatedness of genes or chunks of the genome tells us about is ancestry, and that as a species we are all so closely related to each other that hierarchy or grouping is simply nonexistent, but ancestry can be traced. Ancestry does not equal grouping.

References and Further Reading

In this section we list by chapter the books and journal articles that are directly quoted in the text, and others that provide further information on topics covered in this book. In addition, we list immediately below two books of our own that discuss various issues of race at greater length than has been possible here.

DeSalle, R. and Tattersall, I. (2018). *Troublesome Science: The Use and Misuse of Genetics and Genomics in Understanding Race*. New York: Columbia University Press.

Tattersall, I. and DeSalle, R. (2011). *Race? Debunking a Scientific Myth*. College Station, TX: Texas A&M University Press.

Chapter 1

Darwin, C. (1859). *On the Origin of Species by Natural Selection, or the Preservation of Favoured Races in the Struggle for Life*. London: John Murray.

Dobzhansky, T. (1937). *Genetics and the Origin of Species*. New York: Columbia University Press.

Eldredge, N. and Gould, S. J. (1972). Punctuated equilibria: An alternative to phyletic gradualism. In T. J. M. Schopf (ed.), *Models in Paleobiology*. San Francisco, CA: Freeman Cooper, pp. 82–115.

Fisher, R. A. (1918). The correlation between relatives on the supposition of Mendelian inheritance. *Transactions of the Royal Society of Edinburgh* 52: 399–433.

Huerta-Sánchez, E., Jin, X. Asan, et al. (2014). Altitude adaptation in Tibetans caused by introgression of Denisovan-like DNA. *Nature* 512: 194–197.

Kampourakis, K. (2015). Myth 16: That Gregor Mendel was a lonely pioneer of genetics, being ahead of his time. In R. L. Numbers and K. Kampourakis (eds), *Newton's Apple and Other Myths about Science*. Cambridge, MA: Harvard University Press, pp. 129–138.

Mayr, E. (1942). *Systematics and the Origin of Species, from the Viewpoint of a Zoologist*. Cambridge, MA: Harvard University Press.

Mendel, J. G. (1866). *Versuche über Pflanzenhybriden. Verhandlungen des naturforschenden Vereines in Brünn*, Bd. IV für das Jahr, 1865, Abhandlungen, pp. 3–47.

Morgan, T. H., Sturtevant, A. H., Muller, H. J., and Bridges, C. B. (1915). *The Mechanism of Mendelian Heredity*. New York: Henry Holt.

Olby, R. (1979). Mendel no mendelian? *History of Science* 17(1): 53–72.

Tattersall, I. and DeSalle, R. (2019). *The Accidental* Homo sapiens: *Genetics, Behavior, and Free Will*. New York: Pegasus.

Tattersall, I. (2012). *Masters of the Planet: The Search for Our Human Origins*. New York: Palgrave Macmillan.

Tattersall, I. (2015a). *The Strange Case of the Rickety Cossack, and Other Cautionary Tales from Human Evolution*. New York: Palgrave Macmillan.

Tattersall, I. (2015b). Homo ergaster and its contemporaries. In W. Henke and I. Tattersall (eds), *Handbook of Paleoanthropology*, 2nd ed. Heidelberg: Springer, vol. 3, pp. 2167–2188.

Tattersall, I. (2022). *Understanding Human Evolution*. Cambridge: Cambridge University Press.

Villanea, F. A., and Schraiber, J. G. (2019). Multiple episodes of interbreeding between Neanderthal and modern humans. *Nature Ecology and Evolution* 3: 39–44.

Chapter 2

Agassiz, L. (1850). Diversity of origin of the human races. *Christian Examiner* 49: 110–145.

Bernier, F. (1684). Nouvelle division de la Terre. *Journal des Scavans*, April 14: 133–160.

Blumenbach, J. F. (1775–1795). *De Generis Humane Varietate Nativa*. Trans. T. Bendyshe (1865). Reissue by Elibron Classics.

Buffon, G. L. L., Cte de. (1749 *et seq.*; 1860 trans). *A Natural History. General and Particular; Containing the History and Theory of the Earth, etc.* Excerpted in Count, E. W. (1950). *This Is Race: An Anthology Selected from the International Literature on the Races of Man.*

Gobineau, J. A. de. (1853–1855). *Essai Sur l'Inegalité des Races Humaines*, 4 vols. Trans.: Essays on the Inequality of Human Races. New York: Howard Fertig, 1999.

Kant, I. (1775). *Von der Verschiedenen Racen der Mensche*. Excerpted in Count, E. W. (1950). *This Is Race: An Anthology Selected from the International Literature on the Races of Man*.

Lamarck, J. B. Chevalier de. (1809). *Philosophie Zoologique*. Excerpted in Count, E. W. (1950). *This Is Race: An Anthology Selected from the International Literature on the Races of Man*.

Linnaeus, C. (1758). *Systema Naturae*, 10th ed. Stockholm: Salvius.

Morton, S. (1839). *Crania Americana*. Philadelphia, PA: John Penington.

Nott, J. C. and Gliddon, G. R. (1854). *Types of Mankind: Ethnological Researches*. Philadelphia, PA: Lippincott and Grambo.

Scorrer, K., Faillace, K. E., Hildred, A. et al. (2021). Diversity aboard a Tudor warship: Investigating the origins of the *Mary Rose* crew using multi-isotope analysis. *Royal Society Open Science* 8: 5.

UNESCO. (1950). Statement on race. *International Social Science Bulletin* 3: 154–158.

Chapter 3

AAPA (1996). Statement on biological aspects of race. *American Journal of Physical Anthropology* 101: 569–570.

Brace, C. L. and Montagu, A. S. (1965). *Man's Evolution: An Introduction to Physical Anthropology*. New York: Macmillan.

Coon, C. (1962). *The Origin of Races*. New York: Knopf.

Coon, C. and Hunt, E. E. (1965). *The Living Races of Man*. New York: Knopf.

Count, E. W. (ed.) (1950). *This Is Race: An Anthology Selected from the International Literature on the Races of Man*. New York: Henry Schuman.

Darwin, C. (1871). *The Descent of Man in Relation to Sex*. London: John Murray, 2 vols.

Garn, S. M. (1961). *Human Races*. Springfield, IL: Charles C. Thomas.

Galton, F. (1869). *Hereditary Genius: An Inquiry into Its Laws and Consequences*. London: Macmillan.

Grant, M. (1916). *The Passing of the Great Race*. New York: Scribner.

Klaatsch, H. (1899). Die Stellung des Menschen in der Reihe der Saugetiere, speziell der Primaten und der Modus seiner Herausbildung aus einer nied. *Globus* 76(21): 329–332; 76(22): 354–357.

Haeckel, E. (1868). *Natürlische Schöpfungsgeschichte*, cited from English trans: *The History of Creation*, 6th ed. New York: D. Appleton and Co., 1914.

Huxley, T. H. (1863). *Evidence as to Man's Place in Nature*. London: Williams and Norgate.

Keith, A. (1936). *History from Caves: A New Theory of the Origin of Modern Races of Mankind*. London: British Speleological Association.

Livingstone, F. E. (1962). On the nonexistence of human races. In M. F. Montagu (ed.), *The Concept of Race*. New York: Free Press, pp. 46–60.

Marks, J. (2008). Race: Past, present and future. In B. A. Koenig, S.-J. Lee, and S. Richardson (eds), *Revisiting Race in a Genomic Age*. New Brunswick, NJ: Rutgers University Press, pp. 21–38.

Montagu, A. S. (1942). *Man's Most Dangerous Myth: The Fallacy of Race*. New York: Columbia University Press.

Tattersall, I. and DeSalle, R. (2011). *Race? Debunking a Scientific Myth*. College Station, TX: Texas A&M Press.

Vogt, K. (1864). *Lectures on Man: His Place in Creation, and the History of the Earth*. London: Longman.

Wallace, A. R. (1864). The origin of races and the antiquity of man deduced from the theory of "natural selection." *Journal of the Anthropological Society* 2: clvii–clxxvi.

Washburn, S. (1963). The study of race. *American Anthropologist* 65: 521–531.

Weidenreich, F. (1947). Facts and speculations concerning the origin of Homo sapiens. *American Anthropologist* 49: 187–203.

Weiner, J. S. (1957). Physical anthropology: An appraisal. *American Scientist* 45: 504–509.

Wolpoff, M. and Caspari, R. (1998). *Race and Human Evolution: A Fatal Attraction*. Boulder, CO: Westview Press.

Chapter 4

Cann, R. L., Stoneking, M., and Wilson, A. C. (1987). Mitochondrial DNA and human evolution. *Nature* 325(6099): 31–36.

Cavalli-Sforza, L. L., Menozzi, P., and Piazza, A. (1994). *The History and Geography of Human Genes*. Princeton, NJ: Princeton University Press.

DeSalle, R., Tessler, M., and Rosenfeld, J. (2020). *Phylogenomics: A Primer*. Boca Raton, FL: CRC Press.

DeSalle, R., Schierwater, B., and Hadrys, H. (2017). MtDNA: The small workhorse of evolutionary studies. *Frontiers in Bioscience* 22: 873–887.

DeSalle, R. and Hadrys, H. (2017). Evolutionary biology and mitochondrial genomics: 50,000 mitochondrial DNA genomes and counting. *eLS*. https://doi.org/10.1002/9780470015902.a0027270

Edwards, A. W. F. (2003). Human genetic diversity: Lewontin's fallacy. *BioEssays* 25(8): 798–801.

Harper, P. S. (2008). *A Short History of Medical Genetics*. New York: Oxford University Press.

Havrilla, J. M., Pedersen, B. S., Layer, R. M., and Quinlan, A. R. (2019). A map of constrained coding regions in the human genome. *Nature Genetics* 51 (1): 88–95.

Hirschfeld, L. and Hirschfeld, H. (1919). Serological differences between the blood of different races: The result of researches on the Macedonian front. *Lancet* 2: 675–679.

Knoppers, B. M., Zawati, M. H., and Kirby, E. S. (2012). Sampling populations of humans across the world: ELSI issues. *Annual Review of Genomics and Human Genetics* 13: 395–413.

Lek, M., Karczewski, K. J. Minikel, E. V. et al. (2016). Analysis of protein-coding genetic variation in 60,706 humans. *Nature* 536(7616): 285–291.

Lewontin, R. C. (1974). *The Genetic Basis of Evolutionary Change*. New York: Columbia University Press.

Lewontin, R. C. (1972). The apportionment of human diversity. In *Evolutionary Biology*. New York: Springer, pp. 381–398.

Lott, M. T., Leipzig, J. N., Derbeneva, O. H. et al. (2013). mtDNA variation and analysis using Mitomap and Mitomaster. *Current Protocols in Bioinformatics* 44 (1): 1–23.

Nielsen, R., Akey, J. M., Jakobsson, M. et al. (2017). Tracing the peopling of the world through genomics. *Nature* 541: 302–310.

Paskal, W., Paskal, A. M., Dębski, T., Gryziak, M., and Jaworowski, J. (2018). Aspects of modern biobank activity: Comprehensive review. *Pathology and Oncology Research* 24 (4): 771–785.

Provine, W. B. (2020). *The Origins of Theoretical Population Genetics*. Chicago, IL: University of Chicago Press.

Swede, H., Stone, C. L., and Norwood, A. R. (2007). National population-based biobanks for genetic research. *Genetics in Medicine* 9(3): 141–149.

Vogel, F. and Motulsky, A. G. (2013). *Vogel and Motulsky's Human Genetics: Problems and Approaches*. New York: Springer Science & Business Media.

Chapter 5

Cavalli-Sforza, L. L., Menozzi, P., and Piazza, A. (1994). *The History and Geography of Human Genes*. Princeton, NJ: Princeton University Press (abridged paperback edition).

DeSalle, R. (2018). The paleogenomic revolution: New bearings on human dispersal. *Natural History* 126 (8): 23–26.

Lachance, J., and Tishkoff, S. A. (2013). SNP ascertainment bias in population genetic analyses: Why it is important, and how to correct it. *Bioessays* 35: 780–786.

Petr, M., Hajdinjak, M., Fu, Q. et al. (2020). The evolutionary history of Neanderthal and Denisovan Y chromosomes. *Science* 369: 1653–1656.

Posth, C., Renaud, G., Mittnik, A. et al. (2016). Pleistocene mitochondrial genomes suggest a single major dispersal of non-Africans and a Late Glacial population turnover in Europe. *Current Biology* 26: 827–833.

Roychoudhury, A. K., Roychoudhury, N., and Nei, M. (1988). *Human Polymorphic Genes: World Distribution*. New York: Oxford University Press.

Tills, D., Kopec, A. C., and Tills, R. E. (1983). *The Distribution of the Human Blood Groups and Other Polymorphisms*. Oxford: Oxford University Press.

Chapter 6

1000 Genomes Project Consortium. (2015). A global reference for human genetic variation. *Nature* 526(7571): 68–75.

Comfort, N. (2014). Genetics: Under the skin. *Nature* 513: 306–307.

François, O., Currat, M., Ray, N. et al. (2010). Principal component analysis under population genetic models of range expansion and admixture. *Molecular Biology and Evolution* 27: 1257–1268.

Galbusera, P., Lens, L., Schenck, T., Waiyaki, E., and Matthysen, E. (2000). Genetic variability and gene flow in the globally, critically endangered Taita thrush. *Conservation Genetics* 1: 45–55.

Gopalan, P., Hao, W., Blei, D. M., and Storey, J. D. (2016). Scaling probabilistic models of genetic variation to millions of humans. *Nature Genetics* 48(12): 1587–1590.

Han, E., Carbonetto, P., Curtis, R. E. et al. (2017). Clustering of 770,000 genomes reveals post-colonial population structure of North America. *Nature Communications* 8: 1–12.

Li, J. Z., Absher, D. M., Tang, H. et al. (2008). Worldwide human relationships inferred from genome-wide patterns of variation. *Science* 319(5866): 1100–1104.

Novembre, J. (2016). Pritchard, Stephens, and Donnelly on population structure. *Genetics* 204: 391–393.

Price, A. L., Patterson, N. J., Plenge, R. M. et al. (2006). Principal components analysis corrects for stratification in genome-wide association studies. *Nature Genetics* 38 (8): 904–909.

Pritchard, J. K., Stephens, M., and Donnelly, P. (2000). Inference of population structure using multilocus genotype data. *Genetics* 155: 945–959.

Rosenberg, N. A., Pritchard, J. K., Weber, L. et al. (2002). The genetic structure of human populations. Science 298: 2381–2385 (and technical comment, 2003).

Yudell, M. (2014). *Race Unmasked: Biology and Race in the Twentieth Century*. New York: Columbia University Press.

Chapter 7

DeSalle, R. and Tattersall, I. (2018). What a DNA can (and cannot) tell us about the emergence of language and speech. *Journal of Language Evolution* 3: 1–8.

Fang, H., Hui, Q., Lynch, J. et al. (2019). Harmonizing genetic ancestry and self-identified race/ethnicity in genome-wide association studies. *American Journal of Human Genetics* 105: 763–772.

Fricke-Galindo, I., and Falfán-Valencia, R. (2021). Genetics insight for COVID-19 susceptibility and severity: A review. *Frontiers in Immunology* 12: 1057.

Golestaneh, L., Neugarten, J., Fisher, M. et al. (2020). The association of race and COVID-19 mortality. *EClinicalMedicine* 25: 100455.

Jain, A. K. (2010). Data clustering: 50 years beyond K-means. *Pattern Recognition Letters* 31(8): 651–666.

Kahn, J. (2006). Race, pharmacogenomics, and marketing: Putting BiDil in context. *American Journal of Bioethics* 6: W1–5.

Lee, A. C. K., Alwan, N. A., and Morling, J. R. (2020). COVID19, race and public health. *Public Health* 185: A1–A2.

Lewontin, R. (1998). The evolution of cognition: Questions we will never answer. In *An Invitation to Cognitive Science: Methods, Models, and Conceptual Issues*, vol. 4. Cambridge, MA: MIT Press, pp. 106–132.

Roberts, D. E. (2011). What's wrong with race-based medicine: Genes, drugs, and health disparities. *Minnesota Journal of Science and Technology* 12: 1–21.

Sirugo, G., Williams, S. M., and Tishkoff, S. A. (2019). The missing diversity in human genetic studies. *Cell* 177: 26–31.

Tam, V., Patel, N., Turcotte, M. et al. (2019). Benefits and limitations of genome-wide association studies. *Nature Reviews Genetics* 20: 467–484.

Tattersall, I., and DeSalle, R. (2012). *Race? Debunking a Scientific Myth*. College Station, TX: Texas A&M University Press.

Tattersall, I., and DeSalle, R. (2019). *The Accidental* Homo sapiens*: Genetics, Behavior, and Free Will*. New York: Pegasus Press.

Chapter 8

Barsh, G. S. (2003). What controls variation in human skin color? *PLoS Biology* 1 (1): e27.

Beall, C. M. (1981). Optimal birthweights in Peruvian populations at high and low altitudes. *American Journal of Physical Anthropology* 56: 209–216.

Beall, C. M. (1982). A comparison of chest morphology in high altitude Asian and Andean populations. *Human Biology* 54: 145–163.

Beall, C. M. (2003). High altitude adaptations. *The Lancet* 362: s14–s15.

Beja-Pereira, A., Luikart, G., England, P. R. et al. (2003). Gene–culture coevolution between cattle milk protein genes and human lactase genes. *Nature Genetics* 35: 311–313.

Bersaglieri, T., Sabeti, P. C., Patterson, N. et al. (2004). Genetic signatures of strong recent positive selection at the lactase gene. *American Journal of Human Genetics* 74: 1111–1120.

Chen, F., Welker, F., Shen, C. C. et al. (2019). A late Middle Pleistocene Denisovan mandible from the Tibetan Plateau. *Nature* 569: 409–412.

Gross, M. (2017). Ancient genomes of the Americas. *Current Biology* 28: R1365–R1368.

Hollox, E. (2005). Genetics of lactase persistence: Fresh lessons in the history of milk drinking. *European Journal of Human Genetics* 13: 267–269.

Huerta-Sánchez, E., Jin, X., Asan et al. (2014). Altitude adaptation in Tibetans caused by introgression of Denisovan-like DNA. *Nature* 512: 194–197.

Jablonski, N. (2006). *Skin: A Natural History*. Berkeley, CA: University of California Press.

Lee, S. and Piazza, S. J. (2009). Built for speed: Musculoskeletal structure and sporting ability. *Journal of Experimental Biology* 212: 3700–3707.

Moore, L. G., Niermeyer, S. , and Zamudio, S. (1998). Human adaptation to high altitude: Regional and life-cycle perspectives. *Yearbook of Physical Anthropology* 41: 25–64.

Piltulko, V. V., Nikolsky, P. A., Girya, E. Y. et al. (2004). The Yana RHS Site: Humans in the Arctic before the Last Glacial Maximum. *Science* 303: 52–56.

Swallow, D. M. (2003). Genetics of lactase persistence and lactase intolerance. *Annual Review of Genetics* 37: 197–219.

Willerslev, E., and Meltzer, D. J. (2021). Peopling of the Americas as inferred from ancient genomics. *Nature* 594: 356–364.

Chapter 9

Fuerst, J. (2015). The nature of race: The genealogy of the concept and the biological construct's contemporaneous utility. *Open Behavioral Genetics*. https://doi.org/10.26775/OBG.2015.06.18

Gower, J. C. (1972). Measures of taxonomic distance and their analysis. In J. S. Weiner and J. Huizinga (eds.), *The Assessment of Population Affinities in Man*. Oxford:Clarendon Press, pp. 1–24.

Murray, C. (2021). *Facing Reality: Two Truths about Race in America*. New York: Encounter Books.

Murray, C. (2020). *Human Diversity: The Biology of Gender, Race, and Class*. New York: Twelve.

Spencer, Q. (2015). Philosophy of race meets population genetics. *Studies in History and Philosophy of Science Part C: Studies in History and Philosophy of Biological and Biomedical Sciences* 52: 46–55.

Wade, N. (2015). *A Troublesome Inheritance: Genes, Race and Human History*. New York: Penguin Books.

Figure Credits

1.1   Drawn by Patricia Wynne.

2.1   Redrawn from the public domain by Patricia Wynne.

3.1   Redrawn from the public domain by Patricia Wynne.

4.1   Drawing by Patricia Wynne.

4.2   Drawing by Patricia Wynne.

4.3   Drawing by Patricia Wynne from Cann, R. L., Stoneking, M., and Wilson, A. C. (1987). Mitochondrial DNA and human evolution. Nature 325, (6099): 31–36.

4.4   Drawing by Rob Desalle. Compiled from Knoppers, B. M., Zawati, M. H., and Kirby, E. S. (2012). Sampling populations of humans across the world: ELSI issues. *Annual Review of Genomics and Human Genetics* 13: 395–413 and Swede, H., Stone, C. L., and Norwood, A. R. (2007). National population-based biobanks for genetic research. *Genetics in Medicine* 9(3): 141–149.

4.5   Redrawn by Patricia Wynne from Lek, M., Karczewski, K. J., Minikel, E. V. et al. (2016). "Analysis of protein-coding genetic variation in 60,706 humans." *Nature* 536(7616): 285–291.

4.6   Redrawn from https://macarthurlab.org/2017/02/27/the-genome-aggregation-database-gnomad

5.1   Drawing by Rob Desalle from Cavalli-Sforza, L. L., Menozzi, P., and Piazza, A. (1994). *The History and Geography of Human Genes.* Princeton, NJ: Princeton University Press.

5.2   Drawing by Patricia Wynne from Posth, C., Renaud, G., Mittnik, A. et al. (2016). Pleistocene mitochondrial genomes suggest a single major dispersal of non-Africans and a Late Glacial population turnover in Europe. *Current Biology* 26: 827–833.

5.3   Drawing by Patricia Wynne.

5.5   Drawing by Patricia Wynne.

5.7   Drawing by Patricia Wynne.

5.8   Drawing by Patricia Wynne.

6.1   Drawing by Patricia Wynne.

6.2   Drawing by Patricia Wynne.

6.3   Drawing by Patricia Wynne from Novembre, J. (2016). Pritchard, Stephens, and Donnelly on population structure. *Genetics* 204: 391–393.

6.4   Drawing by Patricia Wynne.

6.5   Drawing by Patricia Wynne from Han, E., Carbonetto, P., Curtis, R. E. et al. (2017). Clustering of 770,000 genomes reveals post-colonial population structure of North America. *Nature Communications* 8: 1–12.

8.1   Drawing by Patricia Wynne from NIH.

Index